UNDERWATER ACOUSTIC DIGITAL SIGNAL PROCESSING AND COMMUNICATION SYSTEMS

Underwater Acoustic Digital Signal Processing and Communication Systems

Edited by

Robert S.H. Istepanian
Brunel University

and

Milica Stojanovic
MIT

KLUWER ACADEMIC PUBLISHERS
BOSTON / DORDRECHT / LONDON

A C.I.P. Catalogue record for this book is available from the Library of Congress.

ISBN 0-7923-7304-9

Published by Kluwer Academic Publishers,
P.O. Box 17, 3300 AA Dordrecht, The Netherlands.

Sold and distributed in North, Central and South America
by Kluwer Academic Publishers,
101 Philip Drive, Norwell, MA 02061, U.S.A.

In all other countries, sold and distributed
by Kluwer Academic Publishers,
P.O. Box 322, 3300 AH Dordrecht, The Netherlands.

Printed on acid-free paper

All Rights Reserved
© 2002 Kluwer Academic Publishers, Boston
No part of this work may be reproduced, stored in a retrieval system, or transmitted
in any form or by any means, electronic, mechanical, photocopying, microfilming, recording
or otherwise, without written permission from the Publisher, with the exception
of any material supplied specifically for the purpose of being entered
and executed on a computer system, for exclusive use by the purchaser of the work.

Printed in the Netherlands.

Contents

Preface vii

List of Contributors ix

1. **High Speed Underwater Acoustic Communications** 1
   M. Stojanovic
   1.1. Channel Characteristics and Sustem Design Principles 2
   1.2. Signal Processing for High Speed Communications 12
   1.3. Areas of Further Development 27
   1.4. Bibliographical Notes 32
   1.5. References 32
2. **Synthetic Aperture Mapping and Imaging** 37
   M.E. Zakharia and J. Châtillon
   2.1. Introduction 37
   2.2. Basics in Sides can Sonar 38
   2.3. Basics in Correlation Sonar 45
   2.4. Ambiguities 47
   2.5. Wideband Synthetic Aperture Sonar Processing 48
   2.6. Trajectory Disturbance 57
   2.7. Autofocusing 61
   2.8. Vernier Processing 66
   2.9. The SAMI Project 68
   2.10. Conclusion 84
   2.11. Acknowledgements 85
   2.12. Notes 85
   2.13. References 85
3. **Integrated Progarammable Underwater Acoustic Biotelemetry System** 89
   R.S.H. Istepanian
   3.1. Introduction 90
   3.2. An Overview of Underwater Biotelemetry Systems 91
   3.3. Underwater Programmable Acoustic Biotelemetry 96
   3.4. General Description and Overall Biotelemetry System 106
   3.5. Conclusion 118
   3.6. References 120
4. **Digital Underwater Voice Communications** 127
   H. Sari and B. Woodward
   4.1. Introduction 127
   4.2. System Design 130
   4.3. Speech Coding 133
   4.4. Encoding and Decoding of Speech Parameters 139
   4.5. Speech Signal Synthesis 140
   4.6. Digital Transmission 142
   4.7. Transmission and Detection of Speech 145

	4.8.	Discussion 149
	4.9.	References 150
5.	**Applications of Neural Networks in Underwater Acoustic Signal processing** .. 167
	Z. Zhaoning	
	5.1.	Introduction 167
	5.2.	Fundamentals of Neural Networks 168
	5.3.	Implementation Issues of Neural Networks 176
	5.4.	Application of Eigen Vector-based DOA Estimation 190
	5.5.	References 201
6.	**Statistical Signal processing of Echo Ensembles** 205
	J.D. Penrose and T. Pauli	
	6.1.	Introduction 205
	6.2.	Monostatic Sounding of Single Point Targets 206
	6.3.	Monte Carlo Simulation 212
	6.4.	Target Strength Estimation from Echo Ensembles 215
	6.5.	A Case Study—The Target Strength of Antarctic Krill 218
	6.6.	References 223
7.	**Advanced Coding for Underwater Communication** 227
	H. Junying, L. Liu, F. Haihong and L. Hong	
	7.1.	Introduction 227
	7.2.	Coding in Underwater Acoustic Communication Systems 228
	7.3.	Time Delay Difference Estimation 232
	7.4.	Underwater Acoustic PDS Communication Model 235
	7.5.	Experimental Results of the System 239
	7.6.	Conclusions 244
	7.7.	References 245
8.	**Three-Dimensional Underwater Acoustical Imaging and Processing** 247
	A. Trucco, M. Palmese, A. Fusiello and V. Murino	
	8.1.	Introduction 247
	8.2.	Data Model 248
	8.3.	Acquisition of 3-D Information 250
	8.4.	Matrix Approach and Real-Time Systems 255
	8.5.	3-D Image Representation 256
	8.6.	Acoustic 3-D Image Processing 257
	8.7.	Segmentation and Reconstruction of Underwater Tubular Structures 260
	8.8.	Conclusion 270
	8.9.	Acknowledgements 271
	8.10.	References 271

Index 275

PREFACE

Underwater acoustic digital signal processing and communications is an area of applied research that has witnessed major advances over the past decade. Rapid developments in this area were made possible by the use of powerful digital signal processors (DSPs) whose speed, computational power and portability allowed efficient implementation of complex signal processing algorithms and experimental demonstration of their performance in a variety of underwater environments. The early results served as a motivation for the development of new and improved signal processing methods for underwater applications, which today range from classical sonar signal processing, to remote control of autonomous underwater vehicles and underwater wireless communications.

This book presents the diverse areas of underwater acoustic signal processing and communication systems through a collection of contributions from prominent researchers in these areas. Their results, both new and those published over the past few years, have been assembled to provide what we hope is a comprehensive overview of the recent developments in the field. The book is intended for a general audience of researchers, engineers and students working in the areas of underwater acoustic signal processing. It requires the reader to have a basic understanding of the digital signal processing concepts. Each topic is treated from a theoretical perspective, followed by practical implementation details. We hope that the book can serve both as a study text and an academic reference.

The topics of the book have been chosen to reflect the emergence of new applications of underwater acoustic signal processing. The book is organised in eight chapters. In Chapter 1, M. Stojanovic introduces the readers to the problem of underwater acoustic communications. Basic characteristics of underwater communication channels and existing communication systems are surveyed, and processing methods required for detection of high-rate communication signals are described. In Chapter 2, M.Zakharia and J.Chatillon address the implementation problems associated with synthetic aperture mapping and imaging systems. The performance of several imaging methodologies in sea trials is presented. A description of a micro-controller based underwater biotelemetry system is presented in Chapter 3 by R. Istepanian. The hardware and software of this underwater monitoring system are described, and results of a SCUBA diver's physiological performance are presented. In Chapter 4, digital underwater voice communications are addressed by H. Sari and B. Woodward. They present a DSP-based voice communication system that includes a speech compression method suitable for transmission over an underwater acoustic channel. In chapter 5, Z. Zhaoning addresses application of neural networks to underwater acoustic signal processing. A review of fundamentals of neural networks is given, followed by the description of implementation architectures relevant for underwater acoustic applications. In chapter 6, J.D. Penrose and T. Pauly describe the process of gaining information about targets from ensembles of backscattered acoustic signals. Target strength estimation is addressed from both a theoretical and an experimental viewpoint. Chapter 7, by H. Junjying, L. Liu, F. Haihong and L. Hong, presents a modulation / coding scheme based on delay estimation in the acoustic channel. The concept is demonstrated through experimental results of shallow water testing. The final chapter, by A. Trucco, M. Palmese, A. Fusiello and V. Murino, is devoted to 3-D underwater acoustic imaging. An acoustic model of the scene to be imaged is developed and a method for segmentation and reconstruction of images is presented in the framework of real-time system requirements.

Finally, we would like to express our thanks and gratitude to all the authors for their excellent contributions. We would also like to thank Mr. Finlay and Ms. Lufting at Kluwer for their effort in publishing this book.

Robert S. H. Istepanian, London, U.K.
Milica Stojanovic, Boston, Massachusetts, U.S.A.

November 2001

LIST OF CONTRIBUTORS

Jacques Chatillon
INRS, Avenue de Bourgone
B27, 54501, VANDOEUVRE
Cedex, France

Andrea Fusiello
Department of Computer Science,
University of Verona,
Italy

Feng Haihong
Underwater Acoustics Institute,
Harbin Engineering University,
Harbin 150001,
P.R. China

Liu Hong
Underwater Acoustics Institute,
Harbin Engineering University,
Harbin 150001,
P.R. China

R.S. H. Istepanian
Department of Electronic & Computer Engineering,
Brunel University, Uxbridge, Middlesex, UB8 3PH,
UK
E-mail: Robert.Istepanian@brunel.ac.uk

Hui Junying
Underwater Acoustics Institute,
Harbin Engineering University,
Harbin 150001,
P.R. China

Liu Li
Underwater Acoustics Institute,
Harbin Engineering University,
Harbin 150001,
P.R. China

Vittorio Murino
Department of Computer Science
University of Verona,
Italy

Maria Palmes
Department of Biophysical and Electronic Engineering,
University of Genova,
Italy

T. Pauly
Australian Antarctic Division,
Channel Hwy., Kingston 7050, Tasmania,
Australia

J.D. Penrose
Centre for Marine Science and Technology,
Curtin University of Technology,
Kent St., Bentley 6102, Western Australia,
Australia

H. Sari
Department of Electronic and Electrical Engineering,
Loughborough University, LE11 3TU,
UK

M. Stojanovic
Massachusetts Institute of Technology
Cambridge, MA, 02139,
USA

Andrea Trucco
Department of Biophysical and Electronic Engineering,
University of Genova,
Italy

B. Woodward
Department of Electronic and Electrical Engineering,
Loughborough University, LE11 3TU,
UK

Manell E. Zhakaria
Ecole Navale/ IRENAV
French Naval Academy/ Underwater Acoustics Group
29360 Brest NAVAL,
France

Zheng Zhaoning
Department of Radio Engineering,
Southeast University, Nanjing,
210018,
P.R.China

1 HIGH-SPEED UNDERWATER ACOUSTIC COMMUNICATIONS

Milica Stojanovic
Massachusetts Institute of Technology

Underwater acoustic communications are a rapidly growing field of research and engineering, driven by the expansion of applications which require underwater data transmission without wired connections. In this chapter, we explore the problems of underwater acoustic communications in three parts. The first part presents an overview of modern applications in underwater data transmission and today's achievements in this area. System requirements are reviewed, and propagation characteristics of underwater acoustic channels are given. It is shown that the majority of underwater acoustic channels are severely band-limited, with signal distortions depending on the link configuration, and ranging from benign to extreme ones caused by time-varying multipath propagation and signal phase variations. Examples of existing systems are given, with emphasis on the methods used for intersymbol interference mitigation. Most of these systems use noncoherent or a differentially coherent signal modulation and detection methods. Phase-coherent detection, which offers better efficiency in bandwidth utilization, is the subject of the second part of this chapter. In this part, the design of high-speed digital communication systems, which rely on powerful equalization and multiple sensor signal processing methods is treated. Theoretical aspects of adaptive multichannel equalization are given, followed by a discussion on adaptive algorithm selection and methods for reducing the receiver complexity. An example of experimental performance analysis is presented, and a DSP implementation of the receiver is described. The concluding part is devoted to future research in the area, which is expected to lead towards the development of high-speed mobile acoustic communication systems and underwater communication networks.

1 CHANNEL CHARACTERISTICS AND SYSTEM DESIGN PRINCIPLES

The need for underwater wireless communications exists in applications such as remote control in off-shore oil industry, pollution monitoring in environmental systems, collection of scientific data recorded at ocean-bottom stations and by unmanned underwater vehicles, speech transmission between divers, and mapping of the ocean floor for objects detection and recovery. Wireless underwater communications can be established by transmission of acoustic waves. Radio waves are of little use because they are severely attenuated, while optical waves suffer from scattering and need high precision in pointing the laser beams. Underwater acoustic communication channels are far from ideal. They have very limited bandwidth, and often cause severe signal dispersion in time and frequency [1]-[5].

Among the first modern underwater communication systems was an underwater telephone, which was developed in the forties in the United States for communication with submarines [2]. This device used a single-sideband (SSB) suppressed carrier amplitude modulation in the frequency range of 8-11 kHz, and it was capable of sending acoustic signals over several kilometers. Today, a new generation of systems is made possible by implementing powerful signal processing and data compression algorithms on digital signal processors (DSPs).

During the past few years, significant advancements have been made in the development of underwater acoustic communication systems, in terms of their operational range and data throughput. Acoustically controlled robots have been used to replace divers in performing maintenance of submerged platforms [9]; high-quality video transmission from the bottom of deepest ocean trenches (6500 km) to a surface ship was established [10]; and data telemetry over horizontal distances in excess of 200 kilometers was demonstrated [19].

The development of efficient communication methods makes new applications possible, which, in turn, impose new requirements on the system performance. Many of the developing applications, both commercial and military, require real-time communication with submarines and autonomous, or unmanned underwater vehicles (AUVs, UUVs). Vehicles, robots and stationary sources on underwater moorings, equipped with oceanographic instruments and cameras, are foreseen to operate together in the future underwater data networks.

System Requirements

In the existing systems, there are usually four kinds of signals that are transmitted: control, telemetry, speech and video signals. The achievable data throughputs, and the reliability of an underwater acoustic communication system, as measured by the bit-error rate, must be determined to suit the bandwidth limitations and distortions of underwater acoustic channels.

Control signals include navigation, status information, and commands for underwater robots, vehicles and submerged instrumentation such as pipeline valves or deep ocean moorings. The data rates up to about 1 kilobit per second (kbps) are sufficient for these operations, but very low bit-error rates may be required [4].

Telemetry data is collected by submerged acoustic instruments such as hydrophones, seismometers, sonars, current-meters, chemical sensors, and it also may include low rate image data. Data rates on the order of one to several tens of kbps are required for these applications. The reliability requirements are not so stringent as for the command signals, and a probability of bit error of $10^{-3} - 10^{-4}$ is acceptable for many of the applications.

Speech signals are transmitted between divers and to a surface station. While the existing, commercially available diver communication systems mostly use analog communications, based on single-sideband modulation of the 3 kHz audio signal, research is advancing in the area of synthetic speech transmission for divers, as digital transmission is expected to provide better reliability. Transmission of digitized speech by linear predictive coding (LPC) methods requires rates on the order of several kbps to achieve close-to-toll quality. The bit error rate tolerance of about 10^{-2} makes it a viable technology for poor quality band-limited underwater channels [12, 13].

Video transmission over underwater acoustic channels requires extremely high compression ratios if an acceptable frame transmission rate is to be achieved. Fortunately, underwater images exhibit low contrast, and preserve satisfactory quality if compressed to few bits per pixel. Compression methods, such as the JPEG (Joint Photographic Experts Group) standard, discrete cosine transform, have been used to transmit 256 × 256 pixel still images with 2 bits per pixel, at transmission rates of about one frame per 10 seconds [10]. Further reduction of the required transmission rate seems to be possible by using dedicated compression algorithms, e.g., the discrete wavelet transform [14]. Current achievements report on the development of algorithms capable of achieving compression ratios on the order of 100:1. On the other hand, underwater acoustic transmission of television-quality monochrome video would require compression ratios higher than 1000:1. Hence, the required bit rates for video transmission range from higher than

10 kbps, possibly up to several hundreds of kbps. Performance requirements are moderate, as images will have satisfactory quality at bit error rates on the order of $10^{-3} - 10^{-4}$.

Channel Characteristics

Unlike in the majority of other communication channels, the use of underwater acoustic resources has not been regulated yet by standards. The available bandwidth and transmission range in an underwater acoustic channels depend on the signal-to-noise ratio which is primarily determined by transmission loss and noise level. System performance and its information throughput depend on the signal distortions caused by reverberation, or multipath propagation. Channel characteristics are time-varying and depend on the system location.

Range and Bandwidth

Transmission loss is caused by energy spreading and sound absorption. While the energy spreading loss depends only on the propagation distance, the absorption loss increases not only with range but also with frequency, thus setting the limit on the available bandwidth [1]. In addition to this nominal transmission loss, the received signal level is influenced by the spatial variability of the underwater acoustic channel, such as the formation of shadow zones. Transmission loss at a particular location can be predicted by many of the propagation modeling techniques [1] with various degrees of accuracy. Spatial dependence of transmission loss imposes particularly severe problems for communication with moving sources or receivers.

Noise observed in the ocean consists of man-made noise and ambient noise. In deep ocean, ambient noise dominates, while near shores and in the presence of shipping activity, man-made noise significantly increases the noise level. Most of the ambient noise sources can be described as having a continuous spectrum and Gaussian statistics [1]. As a first approximation, the ambient noise power spectral density is assumed to decay at 20 dB/decade, both in shallow and deep water, over frequencies of interest to communication systems design.

Frequency-dependent transmission loss and noise determine the relationship between the available range, bandwidth and SNR at the receiver input. This dependence is illustrated in Fig.1 which shows the frequency dependent term of SNR for several transmission ranges. (The SNR is evaluated assuming spherical spreading, absorption according to Thorp and a 20 dB/dec decay of the noise power spectral density[1].) Evidently, this dependence influences the choice of a carrier frequency for the desired transmission range. In addition, it determines the relationship between the available range and frequency band. As a result, underwater acoustic communication links can be classified according to range. For a long-range system, operating over 10-100 km, the bandwidth is limited to few kHz (for a very long distance on the order of 1000 km, the available bandwidth falls below a kHz). A medium-range system operating over 1-10 km has a bandwidth on the order of 10 kHz. A short-range system operates over distances less than a km with bandwidth in excess of 10 kHz, while only at very short distances below about 100 m, more than a hundred kHz of bandwidth may be available.

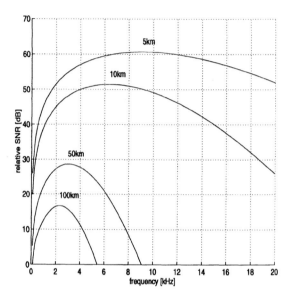

Figure 1: Frequency-dependent portion of SNR.

Multipath

Within a limited bandwidth, the signal is subject to multipath propagation through a channel whose characteristics vary with time and are highly dependent on the location of the transmitter and receiver. In the first place, the multipath spread depends on the link configuration, which is primarily designated as vertical or horizontal. While vertical channels exhibit little time-dispersion, horizontal channels may have extremely long multipath spreads. In a digital communication system which uses a single carrier, multipath propagation causes intersymbol interference (ISI), and an important figure of merit is multipath spread in terms of symbol intervals.

While typical multipath spreads in the commonly used radio channels are on the order of several symbol intervals, in the horizontal underwater acoustic channels they increase to several tens, or a hundred of symbol intervals for moderate to high data rates. For example, a commonly encountered multipath spread of 10 ms in a medium-range shallow water channel causes the ISI to extend over 100 symbols if the system is operating at a rate of 10 kilosymbols per second (ksps).

The mechanisms of multipath formation in the ocean are different in deep and shallow water, and also depend on the frequency and range of transmission. Depending on the system location, there are several typical ways of multipath propagation, determined mostly by the water depth. The definition of shallow and deep water is not a strict one, but usually implies the region of continental shelves, with depth less than about 100 m, and the region past the continental shelves, respectively. One mechanism of multipath formation is by reflections off the bottom, surface and any objects in the water, and this mechanism prevails in shallow water in addition to a possible direct path. Another mechanism, prevalent in deep water, is by ray bending which occurs because the rays of sound tend to reach regions of lower propagation speed. In this way, the sound channel may form by repeated bending of the rays toward the location where sound speed reaches its minimum, called the axis of the deep sound channel. Since there is no loss due to reflections, sound can travel in this way over several thousands of kilometers. Alternatively, the rays bending upwards in deep water may reach the surface focusing in one point where they are reflected, and the process is repeated periodically. The region between two focusing points on the surface is called a convergence zone, and its typical length is 60 -100 km.

The geometry of multipath propagation and its spatial dependence are important for communication systems which use array processing to suppress multipath (e.g., [16], [17]). The design of such systems is often accompanied by the use of a propagation model for predicting the angular distribution of multipath arrivals. Ray theory and the theory of normal modes provide basis for such propagation modeling.

Time-Variation

Associated with each of the deterministic propagation paths (macro-multipaths), which can be modeled accurately, are random signal fluctuations (micro-multipath), which account for the time-variability of the channel response. Some of the random fluctuations can be modeled statistically [1]. These fluctuations include surface scattering due to waves, which is the most important contributor to the overall time variability of the shallow water channel. In deep water, internal waves additionally contribute to the time-variation of the signal propagating along each of deterministic paths.

Surface height displacement can be well modeled as a zero-mean Gaussian random variable, whose power spectrum is completely characterized by the wind speed [1]. Motion of the reflection point results in the Doppler spreading of the surface-reflected signals. Highest Doppler spreads, with values on the order of 10 Hz, are most likely to be found in short and medium range links, which use relatively high frequencies. Note that this effect is present in the channel regardless of the system's mobility.

Statistical channel modeling has significance for communication system design and analysis by simulation. While experimental model-fitting results are limited, short and medium-range channels are often modeled as Rayleigh fading channels. The deep water channel has also been modeled as a Rayleigh fading channel; however, the available measurements are scarce, often making channel modeling a controversial issue [5].

To illustrate the time-varying multipath effects, Figs.2-4 each show an ensemble of channel impulse responses, observed as functions of delay over an interval of time. These responses are estimated from experimental measurements obtained in three typical underwater environments: long-range deep and shallow water, and medium-range shallow water. Relevant system parameters are indicated in the figures.

For a digital communication system which uses adaptive signal processing to track the time-variations of the channel, a relevant parameter is the Doppler spread normalized by the signal bandwidth. This parameter needs to be much less than 1 to enable efficient tracking. Consequently, the implications time-varying multipath bears on the high-speed communication system design are twofold. On one hand, signaling at a high rate causes many adjacent symbols to interfere at the receiver, and requires sophisticated processing to compensate for the ISI. On the other hand, as pulse duration becomes shorter, channel variation over a single symbol interval becomes slower. This allows an adaptive receiver to efficiently track the channel on a symbol-to-symbol basis, or even less frequently, provided, of course, a method for dealing with the resulting time-dispersion.

Examples of Existing Systems

To overcome the difficulties of time-varying multipath dispersion, the design of many underwater acoustic communication systems has so far relied mostly on the use of noncoherent modulation techniques and signaling methods which sacrifice throughput to achieve robustness to channel distortions. Recently, phase-coherent modulation techniques, together with array processing for exploitation of spatial multipath diversity, have been

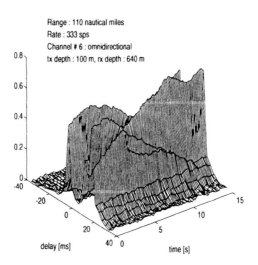

Figure 2: Ensemble of long-range channel responses in deep water (approx. 2000 m) off the coast of California, during the month of January. Carrier frequency is 1 kHz. Range corresponds to three convergence zones. Channel estimates are obtained by recursive least squares estimation using pseudo-random QPSK signals.

shown to provide a feasible means for a more efficient use of the underwater acoustic channel bandwidth. These advancements are expected to result in a new generation of underwater communication systems, with at least an order of magnitude increase in data throughput.

Approaches to system design vary according to the technique used for overcoming the effects of intersymbol interference and signal phase variations. Specifically, these techniques involve the choice of modulation/detection method which provides robustness to the channel impairments, and the choice of transmitter/receiver structure which may include array processing and/or equalization methods. While most of the existing systems operate on the vertical, or the very short-range channels, the systems under development often focus on the severely spread horizontal shallow water channels.

Noncoherent detection of FSK (frequency shift keying) signals has been used to overcome rapid phase variation present in many underwater channels. To deal with the ISI, these systems employ guard times, which are

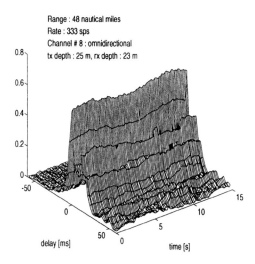

Figure 3: Ensemble of long-range channel responses in shallow water (approx. 50 m) off the coast of New England, during the month of May. Carrier frequency is 1 kHz.

inserted between successive pulses to ensure that all the reverberation vanishes before each subsequent pulse is to be received. The insertion of idle periods of time obviously results in a reduction of the available data throughput. In addition, because fading is correlated among frequencies separated by less than the coherence bandwidth (the inverse of the multipath spread), only those frequency channels with sufficient separation should be used at the same time. This requirement further reduces the system efficiency unless coding is employed so that the adjacent, simultaneously used frequency channels belong to different codewords. A representative noncoherent system [6] uses a multiple FSK modulation technique in the 20-30 kHz band, with maximum bandwidth efficiency of 0.5 bps/Hz. The band is divided into 16 subbands, in each of which a 4-FSK signal is transmitted. This system has successfully been used for telemetry over a 4 km shallow water horizontal path, and a 3 km deep ocean vertical path. It was also used on a less than 1 km long shallow water path, where probabilities of bit error on the order of $10^{-2} - 10^{-3}$ were achieved without coding. The system incorporates the possibility of error correction coding, which improves performance at the expense of reducing the information throughput. The

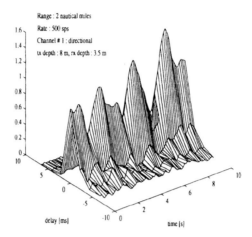

Figure 4: Ensemble of medium-range channel responses in shallow water (approx. 20 m) near the coast of New England, during the month of February. Carrier frequency is 15 kHz.

multiple FSK system is commercially available with a maximum data rate of 1200 bps. This modem has been used in missions of experimental AUVs for vehicle-to-vehicle communication [7].

Despite the low bandwidth efficiency, noncoherent FSK is a good solution for applications where moderate data rates and robust performance are required. Multicarrier modulation techniques in general offer the possibility to eliminate the need for equalization, and as such present an attractive solution for some systems. A system has recently been implemented [8] which uses orthogonal frequency division multiplexing (OFDM) realized with DFT-based filter banks. This system was used on a medium-range channel; however, due to the high frequency separation among the channels (only every fourth channel is used) and relatively long guard times (10 ms guard following a 30 ms pulse), the effective data rate is only 250 bps.

With the goal of increasing the bandwidth efficiency of an underwater acoustic communication system, research focus over the past years has shifted towards phase-coherent modulation techniques, such as PSK (phase shift keying) and QAM (quadrature amplitude modulation).

Depending on the method for carrier synchronization, phase-coherent systems fall into two categories: differentially coherent and purely phase-coherent. The advantage of using differentially encoded PSK (DPSK) with differentially coherent detection is the simple carrier recovery it allows; however, it has a performance loss as compared to coherent detection. Most of the existing systems employ DPSK methods to overcome the problem of carrier phase extraction and tracking. Real-time systems have been implemented mostly for application in vertical and very short range channels, where little multipath is observed and the phase stability is good.

Deep ocean, vertical path channel is used by an image transmission system [10]. This is a 4-DPSK system with carrier frequency of 20 kHz, capable of achieving 16 kbps bottom to surface transmission over 6500 m. The field tests indicate the achievable bit error rates on the order of 10^{-4} with linear equalizer operating under a a least mean squares (LMS) algorithm. Another example of a successfully implemented system for vertical path transmission is that of an underwater image and data transmission system [11]. This system uses a binary DPSK modulation at a rate of 19.2 kbps and a carrier of 53 kHz. The system was used for transmission over 2000 m.

In the very short range channel, where bandwidth in excess of 100 kHz is available, and signal stability is good, phase-coherent system based on 16-QAM has been used [9]. This system operates over 60 m at a carrier frequency of 1 MHz and a data rate of 500 kbps. It is used for communication with an undersea robot which performs maintenance of a submerged platform. A linear equalizer, which uses an LMS algorithm, suffices to reduce the bit error rate from 10^{-4} to 10^{-7} on this channel.

For high-speed communication over longer distances, bandwidth-efficient systems based on phase-coherent signaling methods must allow for considerable ISI in the received signal. These systems employ either some form of array processing, or equalization methods, or a combination thereof, to compensate for the distortions. As an alternative, direct-sequence spread-spectrum has also been used to provide immunity to multipath propagation [15]; however, this technique requires excessive bandwidth.

Array processing has been used both at the transmitter and at the receiver end to isolate a single propagation path and thus eliminate, or at least alleviate the problem of ISI. To excite only a single path of propagation, very large transmitter arrays are required. Instead, the use of parametric sources has been extensively studied [16]. These sources achieve high directivity by relying on the nonlinearity of the medium in the vicinity of a transducer where two or more very high frequencies from the primary projector are mixed highly directive sources rely are mixed, and the resulting signal at difference frequency is transmitted by a virtual array formed in the water column in front of the projector. A major limitation of such a source is in its high power requirements. High directivity implies the problem of pointing errors, and careful positioning is required to ensure absence of multipath. These systems have been employed in short-range shallow water channels where equalization is not deemed feasible due to rapid time-variation of the signal. Binary and quaternary DPSK signals were used achieving data rates of 10 kbps and 20 kbps, respectively, with a carrier frequency of 50 kHz. The estimated bit error rate was on the order $10^{-2} - 10^{-3}$, depending on the range. It was found that this technique is more effective at shorter ranges.

Adaptive beamsteering at the receiver end provides another alternative for reducing signal dispersion in time. The beamformer [17] uses an LMS algorithm to adaptively steer nulls in the direction of a surface reflected wave. The system was tested in shallow water, with DPSK signals transmitted at 10 kbps, and a carrier frequency of 50 kHz, showing a bit error rate of 10^{-2}. Similarly as in the case of a parametric source, it was found that the beamformer encounters difficulties as the range increases relative to depth.

Current state-of-the art in high-speed communications using purely phase-coherent detection of PSK and QAM signals over severely time-spread horizontal channels is based on simultaneous multichannel processing and equalization aided by explicit phase synchronization [18]-[20]. The adaptive multichannel decision-feedback equalizer (DFE) which uses a recursive least squares (RLS) algorithm was tested in a variety of underwater channels, showing satisfactory performance regardless of the link geometry. The achieved data rates of up to 2 kbps over long range channels, and up to 40 kbps over shallow water medium-range channels, are the highest reported to date. In the following section, these methods are discussed in detail.

2 SIGNAL PROCESSING FOR HIGH-SPEED COMMUNICATIONS

In many of the underwater acoustic channels multipath structure may exhibit one or more components which carry the energy similar to that of the principal arrival. As the time progresses, it is not unusual for these components to exceed in energy the principal arrival (e.g., see Fig.2). The fact that the strongest multipath component may not be well defined makes the extraction of carrier reference a difficult task in such a channel.

Receiver Structure

The optimal receiver for multichannel detection in a time-dispersive Gaussian noise channel consists of a bank of filters, one per receiver sensor, each of which is matched to the overall channel response between the transmitter and that sensor [19]. The filter outputs are coherently combined and sampled at the signaling rate. The resulting discrete-time signal represents a set of sufficient statistics for detecting the sequence of transmitted data symbols. The optimal, maximum likelihood sequence estimation (MLSE), even when efficiently implemented using the Viterbi algorithm, has prohibitively high computational complexity to accommodate the ISI span of many of the underwater channels. Linear equalizer on the other hand leads to a computationally simple solution, but lacks the capability to deal with high spectral distortions which are often found in underwater channels. Between these two extreme solutions is the decision-feedback equalizer, which we choose because it offers a good trade-off between performance and complexity [24]. Experimental results have justified this choice.

The optimal receiver gives rise to the structure of an adaptive receiver shown in Fig.5. The input signals to the baseband processor are the A/D converted received signals, translated to baseband using nominal carrier frequency and lowpass filtered. The signals are frame-synchronized prior to any processing. Frame synchronization is accomplished by matched filtering to a known channel probe (a short sequence with good autocorrelation properties). The operations of demodulation and frame synchronization can be performed digitally. A signaling frame is shown in Fig.6. It consists of the channel probe, a pause, and a data block which starts with a training sequence. Frame synchronization is performed at the start of each frame, and it provides coarse alignment in time for the duration of a frame. In a mobile scenario, where bit timing may change significantly over the duration of a frame, the channel probe may be used to obtain a coarse estimate of the Doppler shift based on which the signal is resampled. Resampling is efficiently implemented using polyphase filters [22].

The overall baseband channel response as a function of delay τ at time t can be written as

$$f_k(\tau, t) = h_k(\tau, t) e^{j\theta_k(t)}, \quad k = 1, \ldots K \tag{1}$$

so as to explicitly indicate the carrier phase $\theta_k(t)$ in each of the channels, and the more slowly varying part of the response $h_k(\tau, t)$. The received signal in the kth channel at time t is then modeled as

$$v_k(t) = \sum_n d(n) h_k(t - nT, t) e^{j\theta_k(t)} + \nu_k(t) \tag{2}$$

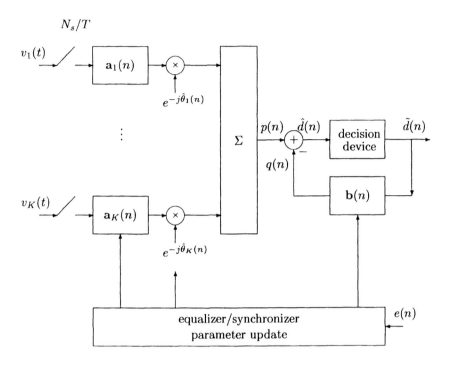

Figure 5: Adaptive multichannel DFE.

where $d(n)$ is the nth transmitted data symbol, and $\nu_k(t)$ is the additive noise.

Knowledge of the channel responses, required by the optimal receiver to implement the matched filters, involves knowledge of the multipath structure, propagation delays and carrier phases. Of these three, carrier phase is the most rapidly changing parameter in the underwater acoustic channel. In classical communication systems, in which carrier and bit synchronization are performed separately from equalization, the presence of strong time-varying multipath affects the performance of a synchronization subsystem, resulting in poor phase tracking capabilities. The residual phase fluctuations on the other hand, impair the equalizer's performance. The equalizer taps, which are complex-valued, may begin to rotate in an attempt to compensate for the residual frequency offset. However, since the

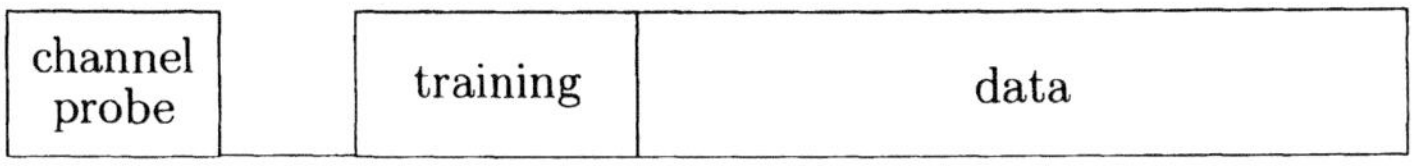

| channel probe | training | data |

Figure 6: Signaling frame.

rate of convergence of the equalizer tap update algorithm is normally lower than the rate at which the carrier phase changes, the convergence is not achieved. A possible solution to this problem is to jointly perform synchronization and equalization, and it appears to be particularly suitable for the underwater acoustic channels with severe multipath. Basic theoretical aspects of simultaneous multiparameter optimization and data detection can be found in [23].

Baseband processing is performed on signals sampled with as few as 2 samples per symbol interval ($N_s = 2$), since the signals are shaped at the transmitter to have maximal frequency less than $1/T$. Raised-cosine spectrum shaping can be used for this purpose, resulting in a maximal frequency of $(1 + \alpha)/2T$, where $\alpha \leq 1$ is the roll-off factor. Since there is no feedback to the analog part of the receiver, the method is suitable for an all-digital implementation.

The sampled signals are processed in the bank of feedforward fractionally spaced equalizers with tap weight vectors denoted by $\mathbf{a}_k$. Sampling is performed starting at an arbitrary time instant, and we assume a sampling rate of $2/T$ without loss of generality. For applications where transmitter and receiver are not moving, but only drifting with water, no explicit adjustment of the sampling clock is needed. It will implicitly be accomplished during the process of adaptive fractionally spaced equalization. The front section of the equalizer will also perform adaptive matched filtering and linear equalization. The output of the feedforward filters is produced once per symbol interval, and since the fractionally spaced equalizers have the capabilities of analog filters they implicitly account for symbol synchronization [24].

Following the feedforward filters is the multichannel carrier phase synchronizer. To correct for the carrier offset, the signals in all channels are phase-shifted by the amount $\hat{\theta}_k$ estimated in the process of joint equalization and synchronization. Depending on the particular channel characteristics, it may not be necessary to have a separate phase-locked loop (PLL) for each of the diversity branches if there is sufficient coherency between the carrier phases in different channels. To accommodate the possibility of large differences in time-varying Doppler frequency shifts observed at different receiver sensors, we consider a general case with as many phase estimates as there are diversity branches.

After coherent combining, the signal is processed by the decision-feedback section of the receiver. A single feedback section is needed because the same information is transmitted in all the channels. The feedback filter with the tap-weight vector **b** forms an estimate of the ISI resulting from the previously transmitted symbols (postcursors), which is then cancelled from the linearly processed signal to form an estimate of the transmitted data symbol, based on which the final decision is made.

This receiver structure is applicable to any linear modulation format, such as M-PSK, or M-QAM, the only difference being in the way in which symbol decision is performed. In addition to combining and equalization, signal processing at the receiver may include the operation of decoding if the signal at the transmitter was encoded.

Parameter Optimization

Having established the receiver structure, we can proceed to determine the optimal values of its parameters. The optimization criterion we use is the minimum mean-squared error (MSE) between the estimated data symbol $\hat{d}(n)$ and the transmitted symbol $d(n)$. The receiver parameters to be determined are the tap weights of the multichannel feedforward equalizer, feedback equalizer coefficients, and the carrier phase estimates. In general, there are two ways of computing the equalizer parameters. One is the direct estimation of the equalizer coefficients driven by the output error, and the other is their computation from the estimated channel impulse response. We shall focus on direct estimation.

Assuming the constant channel impulse response and carrier phase in some short interval of time, one arrives at the optimal values of equalization and synchronization parameters. Let the kth channel feedforward equalizer tap weight vector be

$$\mathbf{a}'_k = [a^k_{-N_1} \cdots a^k_{N_2}]^* \tag{3}$$

where the tap weights are taken as conjugate for convenience of notation. At time nT, associated with the detection of the nth data symbol, the input signal samples stored in the kth feedforward equalizer are represented by the vector

$$\mathbf{v}_k(n) = [v_k(nT + N_1 T/2) \cdots v_k(nT - N_2 T/2)]^T \tag{4}$$

where $N = N_1 + N_2 + 1$ is the number of feedforward taps per channel, and N_1, N_2 are determine to provide the best centering of the signal within the feedforward equalizer, i.e. the highest correlation between the equalizer output and the desired data symbol.

The output of the kth feedforward equalizer, after phase correction by the amount $\hat{\theta}_k$, is given as

$$p_k(n) = \mathbf{a}'_k \mathbf{v}_k(n) e^{-j\hat{\theta}_k} \qquad (5)$$

and the coherent combination of all diversity channels is

$$p(n) = \sum_{i=k}^{K} p_k(n). \qquad (6)$$

The feedback filter coefficients are arranged in a vector

$$\mathbf{b}' = [b_1 \cdots b_M]^* \qquad (7)$$

and the vector of M previous decisions, currently stored in the feedback filter, is denoted as

$$\tilde{\mathbf{d}}(n) = [\tilde{d}(n-1) \cdots \tilde{d}(n-M)]^T. \qquad (8)$$

The output of the feedback filter is now defined as

$$q(n) = \mathbf{b}'\tilde{\mathbf{d}}(n). \qquad (9)$$

The estimate of the data symbol at time n is

$$\hat{d}(n) = p(n) - q(n) \qquad (10)$$

from which the decision $\tilde{d}(n)$ is obtained as the closest signal point. The resulting estimation error is

$$e(n) = d(n) - \hat{d}(n). \qquad (11)$$

The receiver parameters are optimized based on joint minimization of the MSE

$$E = E\{|e^2(n)|\} \qquad (12)$$

with respect to $\{\mathbf{a}_k\}, \mathbf{b}$, and $\{\hat{\theta}_k\}$.

To find the optimal values of the equalizer coefficients, it is convenient to group all the coefficients into a composite vector $\mathbf{c}$, and to express the estimate $\hat{d}(n)$ as

$$\begin{aligned}\hat{d}(n) &= [\mathbf{a}'_1 \cdots \mathbf{a}'_K \ -\mathbf{b}'] \begin{bmatrix} \mathbf{v}_1(n) e^{-j\hat{\theta}_1} \\ \vdots \\ \mathbf{v}_K(n) e^{-j\hat{\theta}_K} \\ \tilde{\mathbf{d}}(n) \end{bmatrix} \\ &= \mathbf{c}'\mathbf{u}(n). \end{aligned} \qquad (13)$$

The MSE can now be expressed as a function of the composite equalizer vector **c**, as

$$\begin{aligned} E &= E\{|d(n) - \mathbf{c}'\mathbf{u}(n)|^2\} \\ &= R_{dd} - 2Re\{\mathbf{c}'\mathbf{R}_{ud}\} + \mathbf{c}'\mathbf{R}_{uu}\mathbf{c} \end{aligned} \quad (14)$$

where we have used the notation $\mathbf{R}_{xy} = E\{\mathbf{x}(n)\mathbf{y}'(n)\}$ for the crosscorrelations. The value of **c** that minimizes the MSE is given by

$$\mathbf{c} = \mathbf{R}_{uu}^{-1}\mathbf{R}_{ud}. \quad (15)$$

The optimal values of the estimates of the carrier phases, $\hat{\theta}_k$, are most easily found if the estimate $\hat{d}(n)$ is represented as

$$\begin{aligned} \hat{d}(n) &= p_k(n) + \sum_{j \neq k} p_j(n) - q(n) \\ &= \mathbf{a}_k'\mathbf{v}_k(n)e^{-j\hat{\theta}_k} + \pi_k(n). \end{aligned} \quad (16)$$

The second term in the last expression is independent of $\hat{\theta}_k$, which makes it possible to express the MSE as

$$\begin{aligned} E &= E\{|d(n) - \pi_k(n) - \mathbf{a}_k'\mathbf{v}_k(n)e^{-j\hat{\theta}_k}|^2\} \\ &= -2Re\{\mathbf{a}_k'E\{\mathbf{v}_k(n)[d(n) - \pi_k(n)]^*\}e^{-j\hat{\theta}_k}\} \\ &+ \text{terms independent of } \hat{\theta}_k \end{aligned} \quad (17)$$

The optimal values $\hat{\theta}_k$ satisfy the gradient equations

$$\frac{\partial E}{\partial \hat{\theta}_k} = -2Im\{\mathbf{a}_k'E\{\mathbf{v}_k(n)[d(n) - \pi_k(n)]^*\}e^{-j\hat{\theta}_k}\} = 0, \quad k = 1, \ldots, K. \quad (18)$$

In order to track the time-varying optimal solution for the receiver parameters, equations (15), (18) should be solved recursively, using updated values of possibly time-varying crosscorrelations. Updating can be carried out continuously, i.e. symbol-by-symbol, or in a block-adaptive manner, in which the parameters are updated only during short training blocks.

The simplest form of an adaptive algorithm is the combination of LMS algorithm for the equalizer coefficients update, and the first-order stochastic gradient update for the digital PLL. Such an algorithm, however, often fails on an underwater acoustic channel, primarily due to the poor phase tracking capabilities. Improved phase tracking capabilities result from the use of a second-order stochastic gradient update for the phase estimates. The instantaneous estimate of the MSE gradient (18) is proportional to

$$\Phi_k(n) = Im\{\mathbf{a}_k'\mathbf{v}_k(n)[d(n) - \pi_k(n)]^*e^{-j\hat{\theta}_k}\} \quad (19)$$

This quantity represents an equivalent of the kth phase detector output. Using the fact that

$$d(n) - \pi_k(n) = p_k(n) + e(n) \qquad (20)$$

the expression (19) is rewritten as

$$\Phi_k(n) = Im\{p_k(n)[p_k(n)+e(n)]^*\} = Im\{p_k(n)e^*(n)\}, \quad k=1,\ldots K. \quad (21)$$

The second-order phase update equations are given by

$$\hat{\theta}_k(n+1) = \hat{\theta}_k(n) + K_{f_1}\Phi_k(n) + K_{f_2}\sum_{m=0}^{n}\Phi_k(m), \quad k=1,\ldots K. \quad (22)$$

It is assumed here that the same proportional and integral tracking constants K_{f_1}, K_{f_2} are used in all diversity channels, and that perfect loop integration is used. Alternative tracking strategies are of course possible. Also, an analogous derivation holds for the case of a single phase estimate for all the channels.

The equalizer coefficients, ideally given by (15), are computed using a suitable adaptive algorithm, driven by the error $e(n)$ and the input $\mathbf{u}(n)$, computed using the current values of the phase estimates. In the training mode, the data symbols $d(n)$ are taken from a known training sequence, while in the decision-directed mode, the values $\tilde{d}(n)$ are used to update the receiver parameters.

Choice of the Adaptive Algorithm

Although the underwater acoustic channels are generally confined to low data rates as compared to many other communication channels, the encountered channel distortions require complex signal processing methods, resulting in high computational load which may exceed the capabilities of the available programmable DSP platforms. Consequently, choosing an efficient adaptive algorithm for receiver implementation, be it for array processing, equalization or both, is an important system design task.

In a majority of recent studies, the LMS-based algorithms are considered due to their low computational complexity, which is linear in the total number of coefficients N [13],[17], [29]. However, the LMS algorithm has a convergence time which may become unacceptably long when large adaptive filters are used (20 N as opposed to 2 N of the RLS algorithm). The total number of coefficients may be very large (more than 100 taps is often needed for spatial and temporal processing in medium and long-range shallow water channels). In addition, the LMS algorithm is very sensitive to the choice of step-size. To overcome this problem, self-optimized LMS algorithms have been used [29], but this results in increased complexity, and increased convergence time.

RLS algorithms, on the other hand, have better convergence properties but higher computational complexity. The quadratic complexity of the standard (form II) RLS algorithm [25] is too high when large adaptive filters need to be implemented. In general, it is desirable that the algorithm be of linear complexity, a property shared by the fast RLS algorithms. A numerically stable fast RLS algorithm [26] has been used for the multichannel equalizer tests in off-line processing, the results of which are shown in Sec.2. A later version of a fast and modular RLS algorithm is given in [27].

Despite its quadratic complexity, a square-root RLS algorithm [28] has been used for real-time implementation because of its excellent numerical stability. The advantage of this algorithm is that it allows the receiver parameters to be updated at arbitrary time instants, rather than every symbol interval, as required by the fast algorithm of [26]. It thus reduces the computational load per each detected symbol. In addition, the updating intervals can be determined adaptively, based on monitoring the mean squared error. Such adaptation methods are especially suitable for use with high transmission rates, where long ISI requires large adaptive filters, but eliminates the need to update the receiver parameters every symbol interval. A different class of adaptive filters, which also have the desired convergence properties and numerical stability, are the lattice filters. RLS lattice algorithms are described in [30].

Reduced-Complexity Receiver Structures

Regardless of the adaptive algorithm used, its computational complexity is proportional to the number of receiver parameters (tap-weights). In addition to focusing on low-complexity algorithms, one may search for a way to reduce the receiver size. Although the use of spatial combining reduces residual ISI and allows shorter equalizers to be used, a broadband combiner may still require a large number of taps to be updated, limiting the practical number of receiving channels to a few. Besides the increase in computational time, very large adaptive filters, which must operate with computationally efficient algorithms, imply increased sensitivity to numerical errors. Unfortunately, some of the fast RLS algorithms [26] preserve numerical stability only at the expense of sacrificing the tracking speed. Another disadvantage of large multichannel equalizers, and perhaps the critical one, lies in their increased noise enhancement, which significantly limits the gain obtained by increasing the number of input channels.

These issues motivate the search for a different multichannel processing strategy in which the size of the adaptive filter will be reduced, but multipath diversity gain preserved. One way to reduce the complexity of the multichannel equalizer of Fig.5 is by the method of pre-combining. Suppose that a large number of input channels K, say more than 10, is available. If each of the K channels is followed by a 50 tap feedforward adaptive filter, the total number of coefficients in the feedforward combining section is more than 500. A reduced-complexity multichannel DFE is shown in Fig.7. It uses a a pre-combiner to reduce a large number of input channels K to a smaller number P for subsequent multichannel equalization. More than one channel at the output of the combiner is usually required if the full diversity gain is to be preserved, but this number, which depends on the type of the transmission channel, is often small.

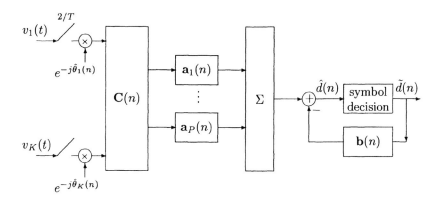

Figure 7: Adaptive multichannel DFE with pre-combining.

The fact that diversity gain may be preserved is explained by multipath correlation across the receiver array. To preserve this processing gain, the $K \times P$ coefficients of the transformation $\mathbf{C}$ are determined jointly with the rest of receiver parameters in a manner analogous to that of Sec.2. The details of receiver optimization can be found in [20]. The adaptive receiver version is based on parallel use of two adaptive algorithms, one for the pre-combiner and one for the P-channel equalizer, where both algorithms are driven by the common error signal $e(n)$. When there is a large number of input channels, and long feedforward equalizers are needed, reduction in the number of adaptively adjusted feedforward taps ($KP + PN$ instead of

KN) can be significant. The choice of P can be made to achieve the least degradation in performance with respect to the full complexity K-channel equalizer, while still keeping the receiver complexity at an acceptable level. We shall illustrate this method shortly by an example.

Another approach in the design of efficient receiver structures is to focus on reducing the number of equalizer taps. A conventional equalizer is designed to span all of the channel response. However, if the channel is characterized by several distinct multipath arrivals separated in time by intervals of negligible reverberation, an equalizer may be designed to have fewer taps. Such method, termed sparse equalization, was applied to detection of QPSK signals recorded in the Arctic waters, for which results show an order of magnitude reduction in computational load [31]. By reducing the number of adaptively adjusted parameters, this approach also makes it possible to use simple updating algorithms, such as standard RLS algorithms which have good numerical stability. Finally, in channels which are naturally sparse, discarding the low-magnitude equalizer taps in fact results in improved performance since no unnecessary noise is processed.

Experimental Performance Analysis

The method of adaptive multichannel combining and equalization was demonstrated to be effective in a variety underwater channels with fundamentally different mechanisms of sound propagation. These channels include the long and medium-range, deep and shallow water channels of Figs.2-4. We shall examine the receiver performance on a long-range shallow water channel.

The experiment we shall describe was conducted by the Woods Hole Oceanographic Institution in the region of New England Continental Shelf, in May of 1992. The transmission ranges were between 15 and 65 nautical miles, and the receiver was positioned in about 50 m deep water with a vertical array of 20 sensors spanning depths from 15 to 35 m. The transmitter power was 193 dB re μPa, and a carrier frequency of 1 kHz was used, with a transducer bandwidth of 1 kHz. At 48 nautical miles (approximately 90 km) an ensemble of estimated channel responses is shown in Fig.3. The channel response consists of a fairly stable main arrival followed by extremely long, unstable multipath. The modulation formats were QPSK, 8-QAM and 8-PSK, and the symbol rates were varied from 1 to 1000 symbols per second.

The signaling frame, as shown in Fig.6, was used with the channel probe consisting of a 13 element Barker code transmitted in phase and in quadrature at the data rate. The signals were shaped using a cosine roll-off filter with roll-off factor α=0.5, digitally implemented by truncating the impulse response to ±2 symbol intervals.

Fig.8 shows the results of signal processing for the transmission rate of 500 sps. The modulation format is 8-PSK, resulting in the bit rate of 1500 bps. Three channels are combined in this example, and they are numbered according to depth, starting from the one closest to the surface. The upper left plot shows the snapshots of the three channel responses, as estimated from the preamble. There is substantial coherence between the channel response magnitudes due to the low separation between the array elements. The ISI is such that the scatter plot of unequalized, symbol rate sampled received signal is completely smeared, even if the signal is phase-synchronized.

The receiver parameters are indicated in the figure: N and M denote the number of feedforward and feedback coefficients, respectively, f.f. denotes the forgetting factor of the RLS algorithm, which accounts for the exponential weighting of the past data, and K_{f_1}, K_{f_2} denote the tracking constants of the PLLs. P_e is the fraction of erroneous decisions in the processed data block. When a single-channel equalizer is used, the scatter plot of estimated data symbols $\hat{d}(n)$ is shown in the upper right corner. Although the single-channel algorithm converges, the performance is not very good, with an estimated probability of error on the order of 10^{-2}. Performance of the multichannel equalizer with $K = 3$ channels is illustrated in the remaining three plots. The estimated MSE indicates steady convergence in decision-directed mode of operation. The number of training symbols required by the RLS algorithm is about twice the total number of adaptively adjusted coefficients. Phase estimates for the three channels are shown together, after each has been scaled to remove the constant frequency-offset term (the average of the Doppler shifts is indicated in the plot). Combining the three channels eliminates all the decision errors and results in about 3 dB better output SNR than in the case of a single-channel equalizer. The multichannel gain normally increases with the number of input channels, but is limited by residual ISI, the noise enhancement in the feedforward equalizers, and the amount of diversity available in the given channel.

As the data rate is increased, these limitations begin to dominate the receiver performance. At 1000 sps, which was the maximum data rate available in the given frequency range, the extent of ISI becomes high enough to prevent successful operation of a single-channel equalizer. A five channel equalizer opens the eye in the output scatter plot, but each feedforward filter needs close to a hundred taps. The reduced-complexity multichannel equalizer offers a solution in this case. Fig.9 shows the result of QPSK signal processing using a $K = 7$ to $P = 3$ pre-combiner followed by a three-channel DFE. Excellent performance, with a clearly open eye and no detected errors is achieved. A very interesting conclusion is also drawn from this study, regarding the receiver performance as a function of the reduced number of channels P. With the total number of input channels fixed to $K = 7$, the reduced number of channels P was varied, and in each case the steady state MSE was observed. With $P = 1$, the performance is poor. The output SNR (the inverse of the MSE) initially increases rapidly with an increase in the number of channels P. However, after a certain value, in this case $P = 3$, the performance saturates, and the SNR retains the same value with further increase in P. In other words, the receiver achieves the full ($K = 7$ in this case) multichannel processing gain at a complexity determined by the number of equalizer channels $P = 3$. The optimal value of P is a parameter inherent to the transmission channel and the system configuration.

Theoretical performance analysis of the adaptive multichannel equalizer requires knowledge of the channel and noise statistics. Under the assumption of a Rayleigh fading channel, an approximate analysis is possible, and it is discussed in [21]. This analysis, as well as the experimental results, confirms the fact that low signaling rates at which ISI may be negligible are not a good choice for rapidly varying channels. On such channels, choosing a higher signaling rate results in better performance through improved tracking capability of the adaptive algorithm.

Related Applications

The principles of multichannel equalization are not limited to the problem of ISI suppression, but apply to a general problem of interference suppression. The sources of interference in underwater acoustic communications include external interference, such as noise coming from on-board machinery or the underwater vehicle launch noise. The internal noise has signal-like characteristics, and it arises in the form of echo in full-duplex systems, and as multiple-access interference in underwater communication networks. In configuration as a noise canceler, the reference of external interfering signal is fed to one of the multichannel receiver inputs, while neither the receiver structure, nor its algorithm change. Performance results with band-limited white noise and multiple sinusoidal interference [32] show the receiver's effectiveness in cancelling the interference while simultaneously detecting the desired signal, which is achieved by virtue of having the training sequence.

A multiple-access communication system represents a special case of structured interference environment. In this situation, two or more sources transmit to a common receiver overlapping in both time and frequency. Since the bandwidth is so scarce in underwater acoustic channels, little processing gain is used to separate the different users' signals by code-division.

The fact that transmission loss varies significantly with range additionally contributes to the near-far effect in the underwater acoustic channels. Two categories of multiuser receivers that have been considered are the centralized receiver, at which the signals of all the users have to be detected (e.g., up-link reception at a surface buoy which serves as a central network node), and the decentralized receiver, at which only the desired user's signal needs to be detected (e.g., down-link reception by an ocean-bottom node). The adaptive multichannel receiver of Fig.7 was experimentally shown to have excellent capabilities in the role of a decentralized multiuser detector, operating without the knowledge of the interfering signal. A multiuser multichannel equalizer, based on the principles of joint equalization, synchronization and multiple-access interference cancellation was successfully tested in experimental trials [33]. Array processing plays a crucial role in the detection of multiuser signals, but is associated with the problem of computational complexity.

Receiver Implementation

Digital technology for underwater acoustic communication systems implementation is a natural choice, having in mind the operating frequencies of these systems. Current converter technology can easily digitize signals which are in the frequency range of interest to underwater acoustic communications, with enough dynamic range to support required receiver selectivity. Another advantage of digital processing is that receiver structures can be easily replicated and tolerance problems associated with analog designs are eliminated. This is of particular importance for operations like array processing which may be delay-sensitive. In addition, digital processing of the signal can compensate for non-ideal operation of the analog front-end.

Programmable DSP platforms currently support operations of more than 100 Mips (mega instructions per second) in fixed-point implementations and comparatively high speeds in floating-point implementations. In addition, multiple DSP boards are commercially available having much greater processing power, and so are specialized processors which support computationally extensive operations specific to communications signal processing.

The advantage of DSP implementation is that it offers great flexibility necessary in early stages of modem development. It also allows easy reconfiguration and changes in the system design (modulation format, equalization algorithm, coding scheme). Small series and the lack of standards also make custom-specific digital circuits very expensive to manufacture for underwater acoustic applications. Although the DSP solutions require more power, this is not a major limitation since various power-reduction techniques, such as putting the transmitter/receiver into sleep mode, can be applied on the system level.

The receiver algorithm described in this section has been implemented on a programmable DSP platform at the Woods Hole Oceanographic Institution. The DSP board, completed in January 1997, has since been under experimental evaluation, in both stationary and mobile communication scenarios. An earlier implementation provided QPSK signaling at 5 kbps, and a carrier frequency of 15 kHz, and was successfully tested in the under-ice shallow water environment [34]. In the current configuration, the modem can support up to eight input channels.

The size of the board is 8 in × 3.5 in × 1.7 in. The board contains eight AD converters, each with sampling capability programmable to up to 200 kHz and 13 bits of resolution. Two output channels have 12-bit DA converters with sampling rate of 500 kHz each.The DSP is a TI320C44, with processing speed of 60 Mflops. The programmable instrument interface includes serial and parallel capabilities and a high-speed data download/upload capability. The memory consists of 6 Mbytes of static RAM and 1 Mbyte of flash RAM. The modem uses a real-time operating system, Acoustic Modem System (AMS), which was developed at the Woods Hole Oceanographic Institution.

In the transmit mode, the modem requires 30 W when coupled to a 180 dBreμPa source. In the receive mode, the power requirement is 8 W, for active reception. Power is supplied by a lithium battery pack, which fits into a cylindrical pressure housing, approximately the size of the board. The modem is powered up into active reception mode by any of a number of wake-up mechanisms. These mechanisms include external activity on the serial port, real-time alarm clock, battery-low signal, or acoustic wake-up by low-power energy detection of a specially encoded FSK sequence. In the passive mode, modem can either be in hibernation, in which it uses only 1 mW of power from a smaller battery and takes about half a second to wake up, or in sleep mode, in which it uses about 20 mW and can wake up instantly. The battery life-time is approximately 2 years in the passive mode, or 2 days of continuous active reception.

Demodulation is performed digitally. Currently, the modem is programmed to use a single second-order PLL for the active channels, with an option for using a multichannel digital PLL. The adaptive filter sizes may be externally determined or chosen autonomously based on the channel mesurements obtained from the Barker probe signal. The adaptive algorithm is a square root RLS, with on-line determination of the updating intervals. The complexity of the algorithm is 9 N^2 per update.

3   AREAS OF FURTHER DEVELOPMENT

At this stage in the development of underwater acoustic communication techniques, with the feasibility of high-rate communications established, a number of research topics emerged which will influence the design of future systems. Some of these topics we have already discussed, namely the reduced-complexity receiver structures and algorithms to enable **efficient receiver implementations**. Beside theoretical advances, future implementations that may require lower power consumption and cost, possibly targeting ASIC (application-specific integrated circuits) realizations of computationally extensive modem functions (coding, equalization) while other will remain in DSP (speech coding, network related functions and functions that require flexibility). Below, we identify several areas of communication engineering in which further developments are likely to provide improved capabilities of future high-speed underwater acoustic communication systems.

Signal Processing

As we have seen at the beginning of this chapter, underwater acoustic communications find application in a variety of autonomous systems. If the adaptive receiver algorithms are to be used in such systems, external assistance in system operation must be minimized. In the first place, receiver initialization involves adjustment of receiver parameters, such as tracking constants in the adaptive algorithm that we have described. These parameters are adjusted according to the instantaneous channel conditions before the actual signal detection can begin. In addition, an increase the noise level, caused for instance by a passing ship, may temporarily disable the communication. The receiver then must wait for the next available training sequence in order to be re-initialized. Two directions are to be pursued towards the development of self-optimized systems. One is further development of **adaptive algorithms** which have the ability to adjust their tracking parameters to fine channel changes. Another is the development of **blind system recovery** techniques, not only for equalization, but also for array processing and synchronization that can operate without a training sequence or a reference signal.

In addition, **speech coding** techniques and **data compression** algorithms suitable for low-contrast underwater images, as well related speech and **image processing** methods, are expected to enable high-rate information transmission over band-limited underwater acoustic channels.

Modulation and Coding

Achieving high throughputs over band-limited channels is conditioned on the use of bandwidth-efficient modulation and coding techniques [24]. Today, results for underwater acoustic communications are confined to signaling schemes whose bandwidth efficiency is at most 3 to 4 bps/Hz. Consequently, there is a need to investigate the performance of higher-level modulation methods. The major concern in doing so is the sensitivity of high-level signal constellations to both phase jitter and time-varying intersymbol interference, which are abundant in the majority of underwater acoustic channels. A possible solution to this problem is the use of special **signal constellation shaping**, which provides more robustness to channel impairments than the conventional PSK or QAM. One method of generating such signals is by differential encoding of both phase and amplitude, which results in high-level constellations that permit differentially coherent demodulation, thus eliminating the need for explicit phase tracking and gain control.

Trellis-coded modulation is well suited for vertical channels which have minimal dispersion. Their use on the horizontal channels requires further investigation because it is associated with the problem of efficient decoding in the presence of long ISI. Namely, the delay in decoding poses problems for an adaptive equalizer which relies on the feedback of instantaneous decisions. Maximum-likelihood sequence estimation, i.e., joint channel estimation and data detection is likely to require implementation of a reduced-complexity sequence estimation algorithm, a number of which is documented in contemporary literature. The design and application of suitable methods for joint equalization and decoding to underwater acoustic channels merits further investigation.

Error correction coding will provide the basis for increasing the reliability of poor-quality underwater acoustic channels. Adequate coding methods, as well as efficient decoding depend on the particular channel and modulation technique used. For example, concatenated codes may be considered for time-varying multipath fading underwater channels.

Channel Modeling and Simulation

Performance analysis of candidate system design methods is subject to the availability of an accurate channel model. While there exists a vast knowledge of both deterministic and statistical modeling of sound propagation underwater, the implications this knowledge bears on the communication channel modeling has only recently received more attention. To date, there is no single statistical channel model widely accepted for any of the underwater acoustic channels. The statistical channel measurements which describe the channel behavior on a time scale of interest to high-speed communication systems are scarce, and focus exclusively on stationary communication scenarios. In a mobile underwater acoustic channel, vehicle speed will be the primary factor determining the time-coherence properties of the channel, and consequently the system design.

Mobile Underwater Communications

The problem of channel variability, already present in applications with a stationary transmitter and receiver, becomes a major limitation for a mobile underwater acoustic communication system. The ratio of the vehicle speed (a fast vehicle moves at several tens of knots, or a nautical miles per hour) to the speed of sound (1500 m/s) exceeds its counterpart in the mobile radio channels by several orders of magnitude, making the problem of time-synchronization very difficult in the underwater acoustic channel. Apart from the carrier phase variation, the mobile underwater acoustic systems will have to deal with the motion-induced pulse compression and dilation. In addition to frame-by-frame signal resampling, to compensate for the motion-induced distortion at high vehicle speeds, algorithms for continuous tracking of the time-varying symbol delay in the presence of underwater multipath are being considered.

Communication Networks

The design of underwater communication networks is characterized by the bandwidth limitation and the long propagation delays in these channels. Due to the bandwidth limitation, frequency-division multiple-access is possible only to a limited extent. Time-division multiple-access, on the other hand, is associated with the problem of efficient protocol design, which arises because of the long propagation delays. As we have already mentioned, a possible solution in such a situation is a system based on code-division multiple-access; however, here again the designer is faced with the problem of low processing gains due to the bandwidth limitation. Consequently, two areas of research are being pursued towards the development of underwater acoustic communication networks. One is the design of network and data link protocols, suited for long propagation delays and strict power requirements encountered in the underwater environment [36]. Another area is that of multiuser detection which allows simultaneous detection of multiple users' signals, thus reducing the need for re-transmissions. Finally, if and when mobile users are enabled to access an underwater network, mobile networking will have to be considered in the light of underwater acoustic channels.

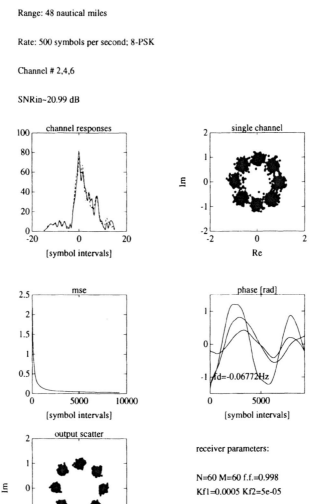

Figure 8: Results of signal processing using adaptive multichannel DFE ($K = 3$). Experimental data from New England Continental Shelf.

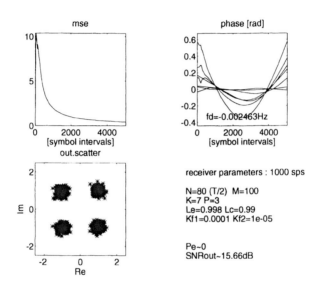

Figure 9: Results of signal processing using reduced-complexity adaptive multichannel DFE ($K = 7, P = 3$). Experimental data from New England Continental Shelf.

BIBLIOGRAPHICAL NOTES

Reference [1] is a classical text which contains a profound treatment of underwater sound propagation and can serve a communications engineer unfamiliar with the propagation characteristics of underwater acoustic channels. Review articles [2]-[5] provide basic concepts of underwater acoustic communications. These articles also contain extensive reference lists for more information on the current state of research and development in the field. Specific examples of communication system design for various applications, such as telemetry, image and speech transmission are given in references [6]-[17], most of which are recent conference publications. Articles [18]-[23] treat system design and performance analysis for high-speed communications based on phase-coherent detection methods. They also contain a number of experimental results. Fundamental principles of digital communications and adaptive filtering are presented in classical texts [24] and [25]. References [26]-[30] contain adaptive algorithms which are of interest to the problems at hand. Finally, references [31]-[36] treat diverse topics in high-speed underwater communications, ranging from efficient modem implementation issues to future underwater data networks.

References

[1] L.Brekhovskikh and Y.Lysanov, *Fundamentals of Ocean Acoustics*, New York: Springer, 1982.

[2] A.Quazi and W.Konrad, "Underwater acoustic communications," *IEEE Comm. Magazine*, pp. 24-29, Mar. 1982.

[3] J.Catipovic, "Performance limitations in underwater acoustic telemetry," *IEEE J. Oceanic Eng.*, vol. 15, pp. 205-216, July 1990.

[4] A.Baggeroer, "Acoustic telemetry - an overview,"*IEEE J. Oceanic Eng.*, vol. 9, pp. 229-235, Oct. 1984.

[5] M.Stojanovic "Recent advances in high rate underwater acoustic communications," *IEEE J. Oceanic Eng.*, pp.125-136, Apr. 1996.

Acoustics," Ph.D. thesis, Northeastern University, Boston, MA, Sept. 1993.

[6] J.Catipovic, M.Deffenbaugh, L.Freitag and D.Frye, "An acoustic telemetry system for deep ocean mooring data acquisition and control," in *Proc. OCEANS'89*, pp. 887-892, Seattle, Washington, Oct. 1989.

[7] S.Chappell et al., "Acoustic communication between two autonomous underwater vehicles," in Proc. *1994 Symposium on AUV Technology*, pp. 462-469, Cambridge, MA, 1994.

[8] S.Coatelan and A.Glavieux, "Design and test of a multicarrier transmission system on the shallow water acoustic channel," in Proc. *OCEANS'94*, pp. III.472-III.477, Brest, France, Sept. 1994.

[9] A.Kaya and S.Yauchi, "An acoustic communication system for subsea robot," in *Proc. OCEANS'89*, pp. 765-770, Seattle, Washington, Oct. 1989.

[10] M.Suzuki and T.Sasaki, "Digital acoustic image transmission system for deep sea research submersible," in *Proc. OCEANS'92*, pp. 567-570, Newport, RI, Oct. 1992.

[11] G.Ayela, M.Nicot and X.Lurton, "New innovative multimodulation acoustic communication system," in Proc. *OCEANS'94*, pp. I.292-I.295, Brest, France, Sept. 1994.

[12] A.Goalic et. al, "Toward a digital acoustic underwater phone," in Proc. *OCEANS'94*, pp. III.489-III.494, Brest, France, Sept. 1994.

[13] B.Woodward and H.Sari, "Digital underwater voice communications," *IEEE J. Oceanic Eng.*, vol. 21, pp.181-192, Apr. 1996.

[14] D.F.Hoag, V.K.Ingle and R.J.Gaudette, "Low-Bit-Rate Coding of Underwater Video Using Wavelet-Based Compression Algorithms," *IEEE J. Oceanic Eng.*, vol. 22, pp.393-400, Apr. 1997.

[15] J.Fischer et al., "A high rate, underwater acoustic data communications transceiver," in *Proc. OCEANS'92*, pp. 571-576, Newport, RI, Oct. 1992.

[16] R.F.W.Coates, M.Zheng and L.Wang, "BASS 300 PARACOM: A "model" underwater parametric communication system," *IEEE J. Oceanic Eng.*, vol. 21, pp.225-232, Apr. 1996.

[17] G.S.Howe et al., "Sub-sea remote communications utilising an adaptive receiving beamformer for multipath suppression," in Proc. *OCEANS'94*, pp. I.313-I.316, Brest, France, Sept. 1994.

[18] M.Stojanovic, J.A.Catipovic and J.G.Proakis, "Phase coherent digital communications for underwater acoustic channels," *IEEE J. Oceanic Eng.*, vol. 19, pp. 100-111, Jan. 1994.

[19] M.Stojanovic, J.A.Catipovic and J.G.Proakis, "Adaptive multichannel combining and equalization for underwater acoustic communications ," *Journal of the Acoustical Society of America*, vol. 94 (3), Pt. 1, pp. 1621-1631, Sept. 1993.

[20] M.Stojanovic, J.A.Catipovic and J.G.Proakis, "Reduced-complexity multichannel processing of underwater acoustic communication signals," *Journal of the Acoustical Society of America*, vol. 98 (2), Pt. 1, pp.961-972, Aug. 1995.

[21] M.Stojanovic, J.G.Proakis and J.A. Catipovic, "Performance of a high rate adaptive equalizer on a shallow water acoustic channel," *J. Acoust. Soc. Amer.*, vol. 100 (4), Pt. 1, pp. 2213-2219, Oct. 1996.

[22] M. Johnson, L. Freitag and M. Stojanovic, "Improved Doppler Tracking and Correction for Underwater Acoustic Communication," in Proc. *ICASSP'97*, vol 1, pp.575-578, Munich, Germany, April, 1997.

[23] H.Kobayashi, "Simultaneous adaptive estimation and decision algorithms for carrier modulated data transmission systems," *IEEE Trans. Comm.* vol. COM-19, pp. 268-280, June 1971.

[24] J.Proakis, *Digital Communications*, New York: McGraw-Hill, 1995.

[25] S.Haykin, *Adaptive Filter Theory*, New Jersey: Prentice Hall 1986.

[26] D.Slock and T.Kailath, "Numerically stable fast transversal filters for recursive least squares adaptive filtering," *IEEE Trans. Sig. Proc.*, vol. SP-39, pp. 92-114, Jan. 1991.

[27] D.Slock, L.Chisci, H.Lev-Ari and T.Kailath, "Modular and numerically stable fast transversal filters for multichannel and multiexperiment RLS," *IEEE Trans. Sig. Proc.*, Vol. 40, pp.784-802, Apr. 1992.

[28] F.Hsu, "Square root Kalman filtering for high-speed data received over fading dispersive HF channels," *IEEE Trans. Inform. Theory*, Vol. IT-28, pp. 753-763, Sept. 1982.

[29] B.Geller, V.Capellano, J.-M.Brossier, A.Essebar and G.Jourdain, "Equalizer for video rate transmission in multipath underwater communications," *IEEE J.Oceanic Eng.*, vol. 21, pp.150-155, Apr. 1996.

[30] F.Ling and J.G.Proakis, "Adaptive lattice decision-feedback equalizers-their performance and application to time-variant multipath channels," *IEEE Trans. Commun.*, vol. 33, pp. 348-356, Apr. 1985.

[31] M.Kocic, D.Brady and M.Stojanovic, "Sparse equalization for real-time digital underwater acoustic communications," in Proc. *OCEANS'95*, San Diego, CA, Oct. 1995.

[32] J.Catipovic, M.Johnson and D.Adams, "Noise cancelling performance of an adaptive receiver for underwater communications," in Proc. *1994 Symposium on AUV Technology*, pp. 171-178, Cambridge, MA, July 1994.

[33] M.Stojanovic and Z.Zvonar, "Multichannel processing of broadband multiuser communication signals in shallow water acoustic channels," *IEEE J. Oceanic Eng.*, pp. 156-166, Apr. 1996.

[34] M.Johnson, D.Herold and J.Catipovic, "The design and performance of a compact underwater acoustic network node," in Proc. *OCEANS'94*, pp. III.467-471, Brest, France, Sept. 94.

[35] M.Johnson, "Utility Acoustic Modem," Technical Report, Woodshole Oceanographic Institution, Jan. 1997.

[36] J.Talavage, T.Thiel and D.Brady, "An efficient store-and-forward protocol for a shallow water acoustic local area network," in Proc. *OCEANS'94*, Brest, France, Sept. 1994.

Chapter 2

Synthetic aperture mapping and imaging

Manell E. Zakharia* and Jacques Châtillon**
*École Navale / IRENAV, French Naval Academy, Underwater Acoustics Group, 29360 Brest NAVAL; France; zakharia@ecole-navale.fr
** INRS, Avenue de Bourgogne BP 27, 54501, VANDOEUVRE cedex, France; chatillon@inrs.fr

1. INTRODUCTION

Synthetic aperture processing for seabed imaging has seen a renewed interest during recent years in both civilian and military applications (see the special issue of the IEEE Journ. of Ocean. Eng. January 1992). Towards the end of the eighties, several prototype systems were developed.

Several experiments conducted with these systems include the wide band CTFM[1] SAS[2] built by Gough et al. [19], the rail experiments conducted by Loggins et al. [27] or by the GESMA[3] [21], the tank experiments achieved at high frequency by Sato et al. [34], by Griffiths et al. [20] or by Sherriff [35], as well as the sea trials carried out by means of the ACID[4] prototype [6] built by the SAMI[5] project team (Synthetic Aperture Mapping and Imaging) [8].

After reviewing some basics on sidescan sonar and correlation sonar, this chapter will present both fundamental work and sea trials results on synthetic aperture mapping of the seabed. Several results issued from the SAMI project will be shown.

2. BASICS IN SIDESCAN SONAR

2.1 Introduction

The operating principle of sidescan sonar is illustrated in figure 2.1.

A towfish containing the physical arrays (transmitter and receiver(s), monostatic sonar operating in the backscattering mode) is towed behind a ship (an AUV[6], or a ROV[7] on a given trajectory. The acoustic observation is obtained by periodic pinging at PRF (Pulse Repetition Frequency) and is perpendicular to the array trajectory. The sound propagates along the slant range axis whilst the arrays travel along the azimuthal axis.

Sonar images are constructed by juxtaposing the intensity of the echoes received from several consecutive pings. It is important to point out, from the very beginning, that the sonar images are quite dissimilar from standard video ones as both axis are from very different nature although they both are expressed in range (or time). One time-scale is the propagation delay of a sound pulse (travelling at about 1500 m/s) and the other one is related to the towfish trajectory (at a few m/s).

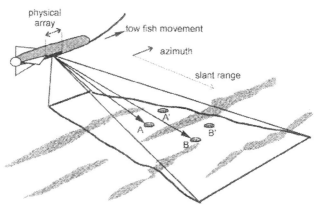

Figure 2.1 Operating principle of sidescan sonar.

2.2 Lambert's law

When the bottom is flat and smooth (figure 2.2a), it acts as a perfect mirror and all the incoming acoustic intensity is reflected in the specular direction: no intensity is backscattered in the transmitter direction. However, when the bottom is rough (with respect to the wavelength), the incoming intensity is scattered in all the directions and part of it is backscattered in the transmitter direction (figure 2.2b). The scattering is governed by the well-known Lambert's law:

$$I_r = \mu I_i (\sin \theta \sin \varphi) dA \qquad (2.1)$$

where: I_i is the incident intensity, I_r is the reflected intensity, θ is the incidence angle and φ the receiving angle; μ is a constant related to the bottom nature (sand, silt, gravel...) and dA is the area of the footprint of the sonar on the seabed.

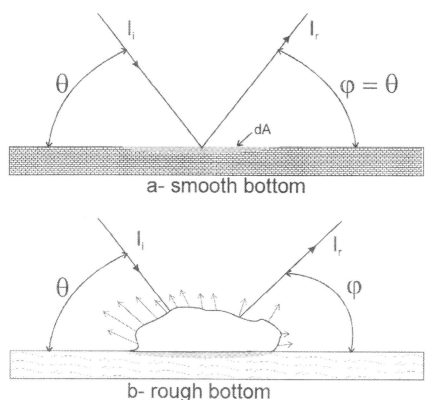

Figure 2.2: Reflection from the seabed.

Apart from the vertical incidence case, the backscattered intensity is thus mainly due to the bottom roughness.

Similar observation can be done on the echo of artificial targets either in the sonar domain (mine, wrecks, containers...) or in the radar domain (planes, tanks, buildings...). A target echo is mainly due to its roughness, edges or irregularities (changing of cross-section): no significant echo is backscattered by plane or (and) smooth surfaces (except in the normal incidence case).

For this reason, the bottom in figure 2.1 is represented by a set of elementary scatterers A, B, A' and B' lying on the bottom. A and B (A' and B') are separated in range while A and A' (B and B ') are separated in azimuth. Although the problem possesses 3D geometry, it is most often intuitively split into two 2D problems: lateral and azimuthal.

2.3 Axial (or lateral) and azimuthal resolutions

Figure 2.3 shows a vertical cross-section of the problem (side or lateral view); it represents a cross-section of the 3D geometry along the y-z plane. Figure 2.4 shows a horizontal one (top or azimuthal view); it corresponds to a projection of the geometry on the x-y plane.

In the following sections, the problem will mainly be addressed as two 2D problems and both cross-sections will be considered separately.

The geometry of the sidescan transducer (linear array) will help in separating both azimuthal and lateral variables: its beam pattern will be omnidirectional in the vertical plane while the horizontal aperture will be as narrow as possible. The dissymmetry of the beam enables to understand the difference in the resolution for both directions.

It is clear, from figure 2.3, that scatterers A and B will have different slant range. Their echoes will possess different time delay. The lateral resolution δ_r is thus a time (range) resolution problem:

$$\delta_r = \delta_t \cdot c \qquad (2.2)$$

where δ_t is the time resolution and c is the sound velocity.

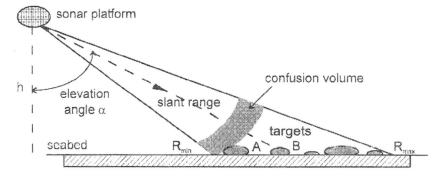

Figure 2.3: Side view.

Figure 2.3 shows that the echoes of all scatterers situated on a spherical cap (or circular one, in the cross-section) will reach the receiver at the same time (even if their elevation angles are different). One can thus define a confusion volume (or surface, in the cross-section) where all the scatterers cannot be separated either because the time resolution is not sufficient or because the angular discrimination of the system is too low. Nevertheless, the discrimination of bottom and volume scatterers is possible for an expert, as they possess different image texture and structure.

The observed area on the seabed is limited by the footprint of the incident beam on the bottom. Only the -3dB main lobe is commonly considered that is limited by R_{min} and R_{max}; [R_{max} -R_{min}] is called the swath.

The axial resolution is illustrated by figure 2.4. The elementary scatterers A and A' are respectively located on the bottom at position y_A and $y_{A'}$ corresponding to lateral range R and R' with:

$$y = R \cdot \sin \theta \tag{2.3}$$

For simplicity, the bottom is assumed to be flat. The distance between A and B ($d = y_A - y_B$), A' and B' ($d' = y_{A'} - y_{B'}$) is the same. In the lateral cross-section, the confusion volume is represented by a circle, of a given "thickness" corresponding to the axial resolution and of a given radius corresponding to the target's range. The slant range for each target corresponds to the intersection of this circle with the bottom. It is clear from the relation above and from the figure that the time delay ($\delta_t = d/c$) between the echo of both scatterer pairs (A-A' and B-B') will increase with their range although the distance between them is constant. The time resolution being constant, the axial resolution on the bottom will improve with range.

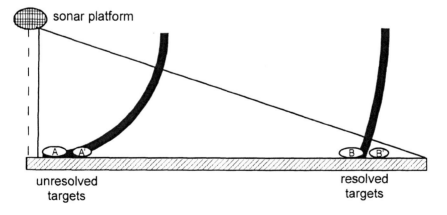

Figure 2.4: Lateral resolution in sidescan sonar.

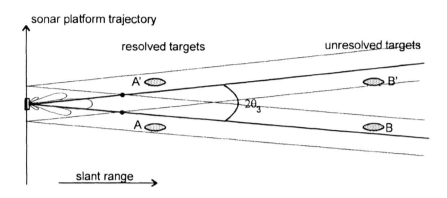

Figure 2.5: Top view, angular resolution in sidescan sonar.

On the contrary, the azimuthal resolution δ_{ra} will decrease with range as illustrated in figure 2.5. This top view shows that the azimuthal resolution is, in fact, an angular resolution (limited, by convention, to the −3 dB beamwidth). The angular resolution being constant (for a given array length), its projection on the seabed (azimuthal cross-section of the footprint) will increase with range and targets A and A' will be resolved whilst targets B and B' will not be resolved although they are separated by the same distance.

The difference between axial and azimuthal resolution illustrates, once again, the heterogeneity of the geometry of sidescan images. Improving the resolution in either direction will require different solutions:
- the improvement of axial resolution will require the improvement of the time resolution (this will lead to the increase of the system bandwidth B; $\delta t \geq 1/B$),
- the improvement of the azimuthal resolution will require the decrease of the horizontal beam of the sonar (i.e. increasing the equivalent length of the array either by increasing the size of the physical array or by considering a "virtual" synthetic array).

2.4 Echoes and shadows

Let us now consider the case of a rough bottom with a prominent bump (seamounts, mines, pipelines, wrecks...) as shown in figure 2.6. While the elementary scatterers lying on the bottom before the bump are illuminated and reflect back acoustic intensity, part of the scatterers behind the bump (marked in black) are located in the shadow of the bump and cannot be reached by the incident wave. At the end of this shadow zone, the elementary scatterers are insonified again and reflect back acoustic intensity.

The received signals will be constituted of several components:
- bottom echo at near vertical incidence (at the very beginning) followed by reverberation from the scatterers, the reverberation level decreases with range due to the Lambert's law;
- echo from the bump; the amplitude of this echo will depend on the slope of the bump in the figure plane but also on its attitude in the plane perpendicular to the figure. In the case of a pipeline, for instance, the amplitude will be low unless the sonar trajectory is parallel to the pipeline;
- shadow zone; the length of this shadow zone is proportional to the height of the bump; the minimum shadow level is limited by the noise level.
- bottom reverberation.

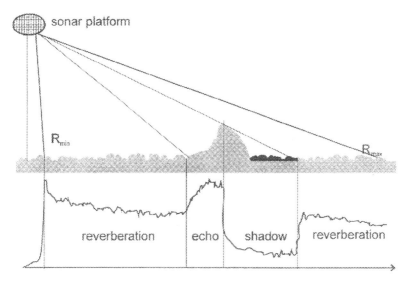

Figure 2.6: Shadow formation mechanisms.

The contrast between the reverberation and the shadow zone is, of course, related to the axial resolution: decreasing this resolution will lead to a smoothing of the echogram leading naturally to a contrast loss. But this contrast is also related to the azimuthal resolution (in the plane perpendicular to figure 2.6. As the confusion volume is constrained between two spherical caps (separated by the range resolution), the shadow in the cross-section constituted by the figure can be corrupted by the reverberation obtained from other cross-sections perpendicular to the cap. Improving the azimuthal resolution will also lead to improving the shadow contrast.

The presence of shadows in sidescan images is a very typical characteristics of these images and the shadow features are most often used for object recognition as they are always present even if the echo is absent (unless the target is flat). Improving both lateral and azimuthal resolution will be needed for the improvement of shadow contrast.

Figure 2.7 shows the 3D geometry of the shadow formation mechanism for a sphere. The diffraction by the edges is neglected and simple optic considerations are used.

Figure 2.8 shows the shadow contours for two simple shape targets (sphere and cylinder). A shadow contour can be easily associated to a given shape: ellipse for a sphere and parallelogram for a cylinder. The shadow distortion due to the elevation angle can be computed by simple geometry considerations.

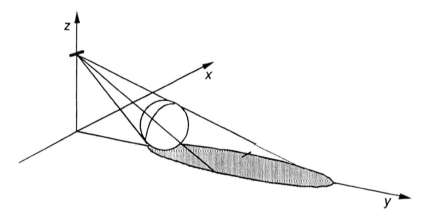

Figure 2.7: 3D geometry of shadow formation mechanism.

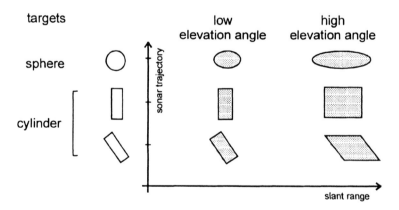

Figure 2.8: Typical shadow of simple shapes.

2.5 Doppler effect

Figure 2.9 shows the position of a point scatterer on the bottom (x-y plane) with respect to the sonar array trajectory (x-axis). The sonar platform movement will induce a Doppler effect leading to a compression or dilatation of the received echoes that has to be considered.

For a constant velocity and a rectilinear trajectory, the towing speed v can be projected on the sonar-target axis and the Doppler effect due to the movement can be computed for each sonar position. The minimum Doppler compression is for R_{min}, the minimum slant range whilst the maximum is obtained for both R_{max} positions.

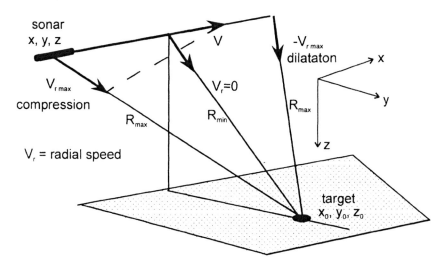

Figure 2.9: Description of "Doppler history" of the echoes.

3. BASICS IN CORRELATION SONAR

Due to the confusion in arrival time between the echoes of all the scatterers situated on a spherical cap, sidescan sonar can only provide "shadow images" and not maps of the seabed. In order to overcome the angular ambiguity a particular effort has to be put on the lateral aspect of the geometry. This is possible thanks to the use of one transmitter and a pair of receivers (swath bathymetry) as illustrated in figure 2.10. The range of a scatterer can be provided by its distance to any receiver and its elevation angle can be obtained by comparing the arrival time between both receivers.

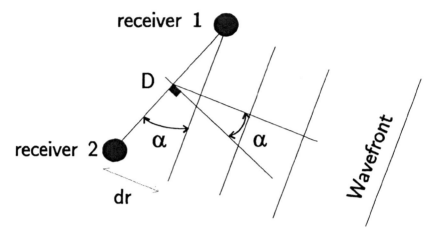

Figure 2.10: Principle of correlation sonar

Let us consider a scatterer at an elevation angle α situated far enough from the receivers so that the reflected wave can be considered as a plane one (due to the size of the elementary scatterers -a few wavelength- this approximation is most often valid). The difference in the acoustic path for the echo to reach both receivers can be approximated by:

$\delta r = D \cdot \sin \alpha$

where D is the distance between both receivers. The corresponding time delay can be expressed as

$\tau = D/c \cdot \sin \alpha$.

This delay can be used for estimating the elevation angle of the scatterer. Given its range and its elevation, the scatterer can thus be accurately positioned in space.

The delay estimation can either be measured by phase difference between the echoes (in the narrow band case) or by time delay (in the wideband case). For improving the accuracy of the time delay estimation, the cross-correlation between both receivers' outputs can be computed:

$$\Gamma(t,\tau) = \int_{t}^{t+\Delta} S_1(t') S_2(t'-\tau) \cdot dt' \qquad (2.4)$$

where S and S are the signals issued from both receivers, Δ is the integration duration, and τ the delay corresponding to a given angle of observation α.

4. AMBIGUITIES

In order to avoid any time and/or angle ambiguity and to provide a proper sampling of the seabed, several conditions are required.

4.1 Range ambiguity

In order to avoid mixing the echoes of two successive transmissions, a new signal transmission cannot occur before all the echoes of the previous transmissions have vanished. Let R_{max} be the maximum range, then:

$$PRF < c \cdot R_{max}/2 \tag{2.5}$$

As R_{max} is defined using the -3dB beamwidth, a security factor of 2 to 4 is commonly used. Using different orthogonal signals for successive transmissions (signal diversity) can loosen this condition.

4.2 Range sampling

For a given azimuthal resolution δ_{ra} at the maximum range R_{max}, the array length will be such as:

$$\delta_{ramax} = 2 \cdot R_{max} \cdot \tan \theta_{3a} \tag{2.6}$$

where θ_{3a} is the -3dB beamwidth (azimuthal plane). The resolution, at the minimum range will then be:

$$\delta_{ramin} = 2 \cdot R_{min} \cdot \tan \theta_{3a} \tag{2.7}$$

$$\delta_{ramin} = \delta_{ramax} \cdot R_{min} / R_{max} \tag{2.8}$$

The azimuthal resolution at close range is excessive unless dynamic focusing techniques are used (i.e. the closer the range, the shorter the receiving array).

In order to avoid any spatial undersampling, the seabed map has to be sampled according to the closer range resolution (figure 2.5) far range will then be oversampled.

This will impose the following condition on the towing speed:

$$v < PRF \cdot \delta_{ramax} \cdot R_{min} / R_{max} \qquad (2.9)$$

4.3 Angular ambiguity

The resemblance between two adjacent reverberating cells (or elementary scatterers) can be so high that the correlation between the two cells on a given receiver can be close to the cross-correlation between the echoes of the same cell on two receivers. The definition of time delay and corresponding angle will become ambiguous. The integration time bas to be optimised in order to avoid such an ambiguity. Unfortunately, no exact inequality can be derived. A compromise has to be found between the correlation length of the scatterers and the integration length ($\Delta \cdot c / 2$). A rule of a thumb is to use a few correlation lengths in the integration one.

5. WIDEBAND SYNTHETIC APERTURE SONAR PROCESSING

5.1 Introduction

As we have seen in the previous section a standard sidescan sonar can be represented by a " stop and go" approach:
- ping in position n,
- recover all the echoes corresponding to transmission n,
- process the echoes and build up the line n of the sonar image,
- go to position n + 1 and
- ping again...

As the sonar movement is much slower than the sound velocity, every ping is providing a "flash" description of the seabed at a given position. The array movement is only considered as a way to scan the azimuthal axis and is used implicitly to build up the sonar image.

Unlike conventional processing synthetic aperture processing will include the array movement explicitly in the image formation mechanism.

5.2 Principle of SAS: optimal processing

Let us consider the geometry described in figure 2.11.1. A sonar platform is moving vertically and insonifying a given target lying on the bottom with a beam aperture of $2\theta_3$. Due to the lack of angular resolution, the target is insonified on several pings (limited to the main beam aperture). For every platform position n, this target is reflecting a given echo. The echoes are located on a trajectory given by the system geometry (2.11.2). For a given position, the slant range of the target is given by:

$$R_n^2 = (x_n - x_0)^2 + (y_n - y_0)^2 + (z_n - z_0)^2 \qquad (2.10)$$

where x_n, y_n, z_n are the array co-ordinates and x_0, y_0, z_0 are the target co-ordinates (figure 2.9).

For a constant speed platform travelling on the x-axis, $x_n = v \cdot t$, $y_n = 0$ and $z_n = 0$, where v is the platform velocity and t is the time.

The echo received at platform position n can be expressed as:

$$E_n(t) = a_n \cdot F(s(\eta_n(t-\tau_n))) \qquad (2.11)$$

where:
- $S(t)$ is the transmitted signal,
- F is a spatial filter due to the directivity pattern of both the target and the array,
- a_n is a normalising constant,
- η_n is the Doppler effect associated to v_n, the projection of the velocity v on the sonar-target axis: $\eta_n = \dfrac{1 - v_n/c}{1 + v_n/c}$ and
- τ_n is the delay corresponding to the slant range R_n: $\tau_n = \dfrac{R_n}{c}$.
-

For a constant velocity on the x-axis, the expression of the echo delay τ_n can be expressed as:

$$\tau_n = \frac{4}{c^2}(At^2 + R_0^2) \qquad (2.12)$$

where: $A = v^2$ and R_0 is the minimum slant range of the target. The "echo trajectory" is a hyperbola.

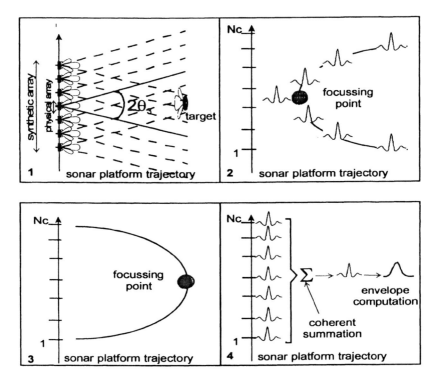

Figure 2.11: Principle of synthetic aperture sonar.

When the elementary target is small enough (with respect to the wavelength) it can be considered as an omnidirectional re-transmitter and its spatial filtering effect can be neglected.

This is not the case of extended (large) targets but, as we mentioned previously, for such targets, only the roughness or cross-section changes will generate significant echo level. Thus, they can be considered, as a first approximation, as a set of omnidirectional "bright spots".

If the directivity pattern of both transmitter and receiver are neglected, the echo expression are simplified:

$$E_n(t) = a_n \cdot s(\eta_n(t - \tau_n)) \qquad (2.13)$$

The synthetic array processing consists in constructing a filter matched in both time and space to the sequence of echo $E_n(t)$ [18].

The impulse response of the matched filter can be written as:

$$SAS(t) = SEQ*(-t) \qquad (2.14)$$

where * notes the conjugate and

$$SEQ = \sum_{N_0-N_c/2}^{N_0+N_c/2} E_n(t) \qquad (2.15)$$

This is the typical synthetic aperture processing used in narrow band radar case where the "Doppler effect" expresses both Doppler compression and echo migration. The processing has to compensate for the phase shift due to both effects. This is the so-called " optimal processing" of the echo sequence.

5.3 Sub-optimal processing

The optimal processing described above leads to a very complex processing architecture: for every target position, the Doppler compression (or shift, in the narrow band case) has to be first computed then compensated before correcting echo migration and synthesising a large synthetic array [7].

If we now neglect the Doppler effect (i.e. $\eta_n \approx 1$ and $a_n = 1$), the expression of the echo sequence can still be simplified:

$$SEQ = \sum_{N_0-N_c/2}^{N_0+N_c/2} S_n((t-\tau_n)) \qquad (2.16)$$

All the echoes of the sequence will have the same expression (for every sonar and target position) apart from their delay that follows a hyperbolic law.

Synthetic aperture processing will then consist in compensating echo delay (figure 2.11.3), adding all the echoes and computing the output envelope (figure 2.11.4). This processing has to be preceded by a filter matched to a single transmitted echo (standard matched filter for pulse compression) .This sub-optimal processing is described in figure 2.12.

Apart from pulse compression filter, the SAS processing is now reduced to a simple compensation of echo migration delays. At this stage, it is important to summarise al the approximations that lead to the processing simplification:
- the directivity pattern of the physical arrays (transmitter and receiver) are neglected and
- the directivity pattern of the target and the Doppler compression are neglected.

This last approximation is the major difference between narrow band radar processing and wideband sonar processing.

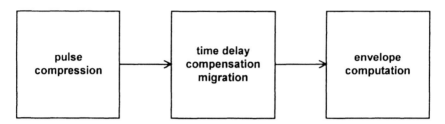

Figure 2.12: Processing Scheme.

As the towing speed v is very small compared to the sound velocity c, the Doppler rate can be approximated by:

$$\eta_n = \frac{1 - v_n/c}{1 + v_n/c} \approx 1 - 2v_n/c \tag{2.17}$$

For a towing speed of 10 m/s (20 knots), the projection on the sonar-target axis can reach, for instance, the value of 7.5 m/s for an angle of 45 degrees. The maximum compression (or dilatation) rate is 1.002 (0.2 %). Although this rate is very low, it can create enough phase shift and/or bias in the echo arrival time that could deteriorate the performance of the processing. This can be avoided by an appropriate signal design.

The spectrum of signals possessing a logarithmic group delay expression is the following:

$$S(f) = A(f) \cdot \exp(i\beta \ln f) \cdot U(f) \tag{2.18}$$

where f is the frequency, A(f) is the spectrum magnitude, U(f) the Heaviside function, and β a real constant.

Their ambiguity function (which expresses the correlation between compressed and original copy of the signal) can be expressed as [42]:

$$\chi(\tau, \eta) = a'(t - \tau) \exp(i\beta' \eta) \tag{2.19}$$

where τ is the propagation delay corresponding to the target position at $t = 0$, a' the envelope of the ambiguity function and β' a phase factor due to the Doppler effect.

Several investigations have shown that an appropriate choice of the time reference (within the transmitted signal) could cancel the bias due to the Doppler for any signal [28]. The selected family is, so-called, "naturally tolerant" to Doppler, as the maximum of the envelope of the ambiguity function is always situated at zero (independently from η). Signal design will then consist in reducing amplitude loss due to the Doppler compression. It leads to the use of Linear Period Modulated (LPM) chirps. In this last family, particular signals have been used that could be expressed in a similar way in both the time and the frequency domain under a stationary phase approximation [2,3], as identified in the following equation:

$$S(f) = A \cdot \exp\left[-\frac{\ln^2(f/f_0)}{\ln(g)}\right] \cdot \cos\left[2\pi b \frac{\ln(f/f_0)}{\ln(g)}\right] \qquad (2.20)$$

where f_0 is the central frequency, and the parameters g and b are related respectively to the bandwidth and to the duration of the signal.

In addition, such signals possess interesting properties of tolerance to acceleration [42] which become of interest when the towing speed varies.

The only remaining effect of the differential Doppler (compression rate variation from one echo to the other due to the geometry of the problem) is a phase shift of the output of the matched filter.

Optimal azimuthal processing should compensate for this shift before any echo stacking. In sub-optimal processing, this compensation can be omitted in order to reduce the processing requirement. Several simulations, using the cited signals, have shown that, for a "reasonably" wide bandwidth (one octave) and a "reasonably" low speed (<12 knots), this differential effect can be neglected [13]. Thanks to this, a "stop-and-go" approach can then be used and only the geometry of the stacking has to be considered.

Several complementary simulations have shown that no visible degradation could be observed due to this last approximation up to towing speed of 30 knots [13].

Under these approximations and when neglecting the influence of the physical array, the sampled expression of the beampattern of the synthetic array can be approximated, at the central frequency, by [7]:

$$\Omega_{sa}(n\Delta) = \frac{\sin\left[2\pi \frac{N_c \Delta}{\lambda_0 R_0} n\Delta\right]}{N \sin\left[2\pi \frac{N_c \Delta}{\lambda_0 R_0} n\Delta\right]} \qquad (2.21)$$

where Δ is the sampling interval on the x-axis, n is the platform position number, N_c is the number of platform positions used for processing and R_0 is the target range.

5.4 Resolution

The angular aperture (at -3dB) of the sampled array obtained is equal to:

$$\theta_{sa} = \frac{\lambda_0}{2L_{sa}} \tag{2.22}$$

where λ_0 is the wavelength corresponding to the central frequency and L_{sa} is the length of the synthetic array.

In order to avoid the range-sampling problem described for standard sidescan sonar, dynamic focusing can be used and the length of the synthetic array can be adapted to the target range. The processing is commonly restricted to the -3dB main lobe of the physical array, then, in case of a line T/R array, the length of the synthetic one is [17]:

$$L_{sa} = R_0 \theta_3 \approx R_0 \frac{\lambda_0}{L_r} \tag{2.23}$$

where R_0 is the minimum target range and L_r is the length of the physical array. The azimuthal resolution of the synthetic array is thus constant and equal to half the physical one:

$$\delta_{sa} = \frac{L_r}{2} \tag{2.24}$$

This simple expression is very important and shows a major difference between standard and synthetic aperture: for a standard aperture, the longer the physical array, the better the resolution whilst, for a synthetic one, the shorter the physical array, the better the resolution.

It is important to note that performance computation has been achieved for the central frequency of the sonar. Several approaches for wideband processing has been investigated (sub-band processing, time domain, frequency domain) and have shown that for a slight increase of processing complexity (1.5 with respect to the narrow band case), the time compensation can be the appropriate solution [9].

The array gain (in dB) due to synthetic aperture processing is range dependent:

$$G = 10 \log \frac{2 R_0 \lambda_0}{L_r^2} \tag{2.25}$$

5.5 Ambiguities

The range ambiguity expressed for standard sidescan (equation 2.5) is still valid while the range sampling constraint (equation 2.9) becomes different due to the use of dynamic focusing. The synthetic aperture processing provides high resolution images with a resolution of Lr/2.

In order to sample the images properly, the synthetic image has to be sampled at this rate; i.e. between two successive pings, the towed fish should not move more than Lr/2. This leads to an extra constraint on the ping rate:

$$v < PRF \cdot \frac{L_r}{2} \qquad (2.26)$$

The problem of ambiguities of synthetic aperture processing has been widely discussed in the literature [25, 30, 32, 33].

The following table shows a comparison of ambiguities and sampling constraints for both conventional sidescan sonar and synthetic aperture one.

system	conventional	SAS	comments
Range ambiguity	$PRF < \dfrac{c}{2R_{max}}$	$PRF < \dfrac{c}{2R_{max}}$	same
Sampling	$v < PRF \cdot \dfrac{\delta_{ra\,max} \cdot R_{min}}{R_{max}}$	$v < PRF \cdot \dfrac{L_r}{R}$	Note 1
Angle ambiguity	A few Γ_s	A few Γ_s	Note 2

- Note 1: the sampling has to be matched to the best resolution; the highest the resolution, the highest the sampling rate; this may lead to misleading interpretation such as: "the SAS requires more severe sampling conditions" which may only be true because SAS possesses higher resolution. For a given resolution, SAS and standard system will require the same sampling constraints.
- Note 2: Γ_s is the correlation length; the volume ambiguity is a 3D ambiguity; in SAS, the volume is reduced by reducing the size of the azimuthal resolution cell, the angular ambiguity constraint is thus also reduced.

5.6 Some noise considerations

As far as noise is considered, it can be assumed that the noise is independent from ping to ping. The synthetic aperture processing will then consist in adding coherent echoes of the same target buried in incoherent noises. As a side effect, the signal to noise will be enhanced by a factor $\sqrt{N_c}$ where N_c is the number of platform positions used for synthetic aperture processing. The improvement of signal to noise ratio will be:

$$\left. S/N \right|_{out} = \left. S/N \right|_{in} + 10 \cdot \log N_c \qquad (2.27)$$

where $\left. S/N \right|_{out}$ is the signal to noise after SAS and $\left. S/N \right|_{in}$ is the one at the SAS processing input.

5.7 A simple example

Figure 2.13 shows the result of processing for two simulated point targets (the very simple case). The compensation of migration (figure 2.11) and the gain of resolution can clearly be seen as well as the gain in signal-to-noise ratio. It can also be noted that the lateral resolution is not affected by synthetic aperture processing.

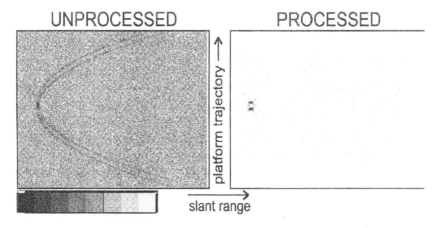

Figure 2.13: Processing example for a point target.
Scale: horizontal [353-358 m], vertical: [0-100 m.], 3 dB/grey level, maximum in black. Input signal-to-noise ratio: 12 dB. Target separation: 4 metres, central frequency: 8 kHz, bandwidth: one octave, physical array: $L_r = 0.4$ m, synthetic array: $L_{sa} = 32$ m.

6. TRAJECTORY DISTURBANCE

6.1 Introduction

The synthetic aperture processing is based on the explicit integration of the sonar platform trajectory in the array beamforming. Having said that, any mismatch between the assumed trajectory and the actual one will lead to the degradation of the sonar image quality. In this section we will study the influence of trajectory disturbances on the quality of the sonar image.

6.2 Definitions

When an appropriate design of the towfish and of the towing system has been achieved, the trajectory disturbances will be rather small and the actual trajectory can be considered as a rectilinear uniform trajectory with some small trajectory disturbances. The towfish possesses 6 degrees of freedom (figure 2.14):
- 3 rotations: yaw, pitch and roll and
- 3 translations: surge, heave and sway.

Any disturbance may be either a periodic one (mechanical resonance of the tow cable, swell, waves...) or a random one (turbulence, currents, microstructures...). As the synthetic aperture processing is based on time delay compensation, the disturbances leading to echo displacement (translations) will be the most important ones. The total disturbance will be the resultant of all the three translations.

The amplitude of the resultant disturbance (figure 2.15) bas to be compared to the central wavelength (λ_0) while its spatial frequency content has to be compared to the length of the synthetic aperture array (L_{sa}).

6.3 some examples of the influence of trajectory disturbances

Figures 2.16 and 2.17 show some example of the influence of the trajectory disturbances on the sonar images. In both cases, the echoes of two point targets are simulated with periodic disturbances and processed as if the trajectory was a perfect one. This will allow the study of the robustness of the processing with respect to the disturbances [14] .In the first case (figure 2.16), the spatial frequency of the disturbances is fixed ($L_{sa}/8$) and the amplitude varies from λ_0 to $\lambda_0/16$. While the image degradation increases with the amplitude, a local observation (such as resolution measurement) is not sufficient as the degradation may be out of the region close to the targets. In the second case (figure 2.17), the amplitude of the degradation is kept constant ($\lambda_0/4$) while its spatial period varies from $L_{sa}/16$ to $2L_{sa}$. These images show that the image degradation depends highly on the period of the disturbances.

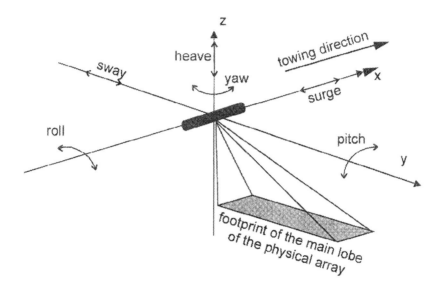

Figure 2.14: Disturbance of the sonar trajectory.

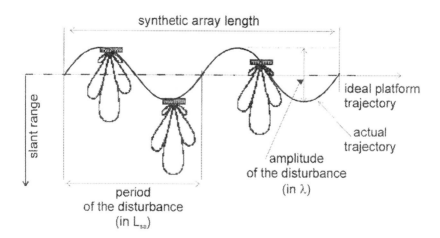

Figure 2.15: Resultant trajectory disturbance.

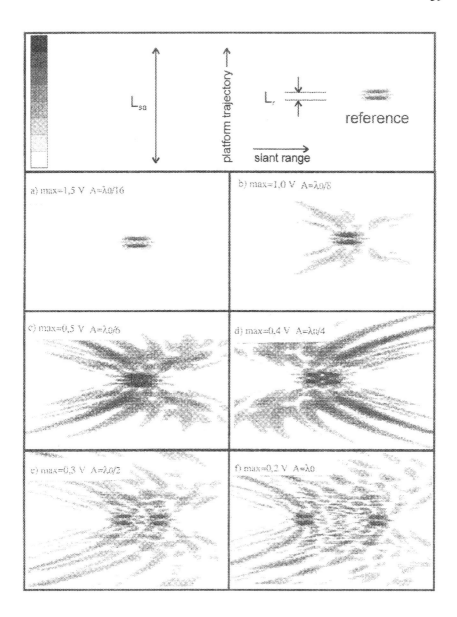

Figure 2.16: Effect of trajectory disturbance for a given spatial period ($L_{sa}/2$) and various amplitudes. Scale: horizontal [245-255 m.], vertical: [0-40 m], 3 dB /grey level, maximum in black.

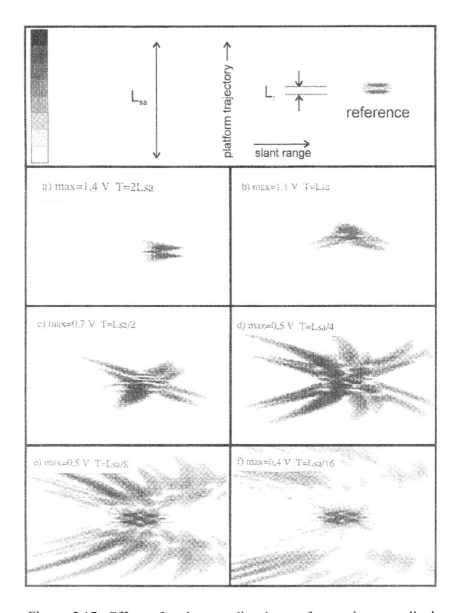

Figure 2.17: Effect of trajectory disturbance for a given amplitude ($\lambda_0 / 4$) and various spatial periods. Scale: horizontal [245-255 m], vertical: [0-40 m], 3 dB /grey level, maximum in black.

6.4 Definition of image quality

The results displayed in both figures 2.16 and 2.17 show that the most intuitive and objective criterion for image quality -the resolution- is not sufficient to determine the image quality. An important visual degradation can be observed while the resolution is only slightly deteriorated. Both local and global criteria are needed for this purpose [12].

After investigating several criteria and comparing to sonar operators definitions, the variance of the normalised image, also called contrast function C(P) has been used:

$$C(P) = \frac{E[I(P,x,y)^2] - E[I(P,x,y)]^2}{E[I(P,x,y)]} = \frac{\sigma_i^2(P)}{E[I(P,x,y)]} \qquad (2.28)$$

where:
- P is the set of parameters defining the disturbances,
- $I(P,x,y)$ is the image obtained with this set of parameters,
- $\sigma_i^2(P)$ is the image variance for this set and
- E stands for expected value.

The image contrast for a given set of parameters will be compared to the contrast obtained with no disturbance. Comparisons have been done using a single point target.

The contrast function variations are given in figures 2.18 a and b for a wideband configuration (one octave) for various periods and amplitudes. Some visual observations of the degradation (figures 2.16 and 2.17) are marked by circles on the curves. Figures 2.18 c and d show similar results for both random and periodic disturbances. One can note that orders of magnitude are comparable.

7. AUTOFOCUSING

7.1 Introduction

The main obstacle to the development of synthetic aperture sonar techniques emphasised in the literature is the loss of echo coherence due to navigation inaccuracies and propagation fluctuations. A great amount of work has been achieved for finding solutions to this problem, in both Synthetic Aperture Radar [5,16] and SAS domains [4, 5, 10, 15, 16, 22, 23].

Synthetic aperture sonar development has to focus on designing a tow fish that could be as stable as possible at any towing speeds. Trajectory disturbances may be measured by a positioning system such an inertial navigation unit. Given the possible image degradations shown in the previous paragraph, one can see that, even for a low frequency sonar system (a few kHz) an accuracy of a few centimetres over a distance of a few tens of meters may be required. Such accuracy cannot be obtained with a reasonable cost navigation unit. Additional methods are needed for correcting the trajectory disturbances.

Several methods using the redundancy between consecutive pings and/or sub-arrays have been developed in both the sonar and the radar domains. Two methods will be developed in this chapter: the bright spot approach and the DPC (Differential Phase Shift).

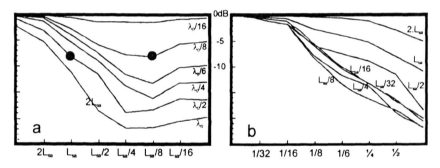

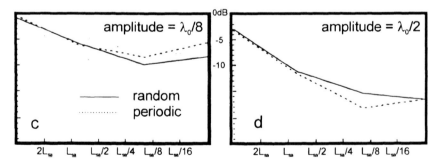

Figure 2.18: Contrast degradation due to trajectory disturbances

7.2 Bright spots method

The algorithm described in this paragraph is based on the coherence of the echoes of a point scatterer given some assumptions about the sonar trajectory: the trajectory is assumed to be relatively stable with additive disturbances being small compared to the echo migration.

For a perfect trajectory, the migration law of the echoes is a hyperbola) as mentioned previously (equation 2.12):

$$\tau_n^2 = \frac{4}{c^2}(At^2 + R_0^2)$$

For a disturbed trajectory) the theoretical echo delays (τ_n) will be affected by an additional random error (ε_n):

$$\tilde{\tau}_n = \tau_n + \varepsilon_n \qquad (2.29)$$

The actual times of arrival ($\tilde{\tau}_n$) are no longer matched to the spatial filter and, in this case, conventional focusing could be spoiled by a significant loss of resolution (as shown in the previous paragraph).

The auto-focusing algorithm consists in tracking the echoes of a given bright spot on the bottom. It operates into several steps (see figure 2.19):

- computing the arrival time of the echoes,
- finding the best fit hyperbola (polynomial approximation using a root mean square minimisation),
- estimating the position of the centre of the hyperbola (n_0, τ_0) and the mean speed of the sonar during the survey of the point target,
- computing the delay fluctuations (error with respect to the mean hyperbola),
- correcting the delay fluctuations on the raw data and
- computing synthetic aperture using the corrected data, the estimated centre position and the estimated hyperbola.

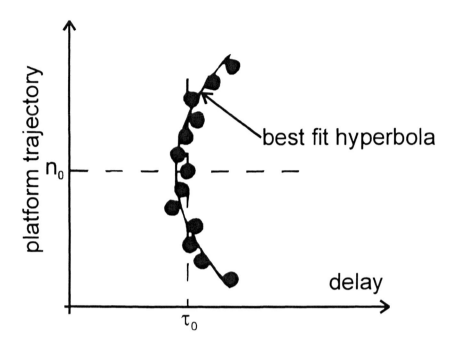

Figure 2.19: Estimation of echo arrival time for autofocusing

7.3 DPC method

The " Displaced Phase centres" method can be considered as one of the general methods using the redundancy of information (thanks to the use of sub-arrays) for compensating echo delay (or phase) fluctuations due to the random movements of the platform and to medium fluctuations. In the early eighties Raven [29] patented the method, and Sheriff [35] was the first one to apply the method to synthetic aperture sonar.

The advantage of such a method is that no predominant scatterer is required to estimate the navigation errors, which allows the processing to be used in the case of reverberating seabed.

The technique consists in processing cross-correlation functions between pairs of elements (sub-arrays) in order to estimate the along-track unknown movement of the platform during two successive transmissions. The best matched elements are then used to beamform the array, for the two successive positions, on a given portion of the seabed. The cross-correlation of the beams obtained by this beamforming permits finally to estimate time (or phase) corrections with increased accuracy and robustness.

Let us consider the physical array composed of one transmitter and several receiving sub-arrays. One can define the "displaced phase centre" (DPC) as the equivalent position of a transmitter/receiver (T/R) which would receive the echo from the target with the same way-and-return range that the sum of the one-way ranges between the transmit element and the receive element. Figure 2.20 (top) shows the geometry of this virtual element DPC.

The bottom part of the figure shows how two DPC can be matched between two successive transmissions, by displaying a complex array (one transmitter, 3 receivers) moving from one position to another. This array permits to define 3 DPC centres, between the transmitter T and each receiver 1, 2 and 3. Although the array has moved (translation and rotation) between ping 1 (white) and ping 2 (grey), the DPC-T/R_2 is, for ping 1, exactly at the same position (located by the fixed reference mark) than the DPC-T/R_1 for ping 2.

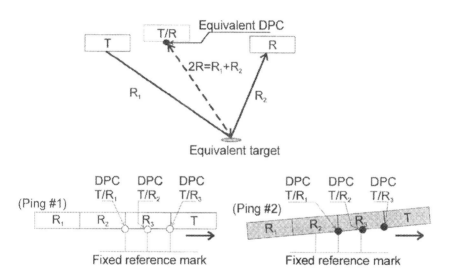

Figure 2.20: Principle of DPC autofocusing method.

This exact superimposition enables the best correlation, between the echoes of T/R_2 for ping 1 and the echoes of T/R_1 for ping 2.

The figure shows also that DPC-T/R_2, for ping 2, is not too far from the position of DPC-T/R_3 for ping 1. The cross-correlation can then have a value indicating this proximity in space. These workable values can then indicate what is the displacement of the array between ping 1 and ping 2 for two points in space, which permits to compensate for residual across-track motion and azimuthal rotation.

Some results have been published by Sheriff [35] for simulations and tank data, and Tonard [36] for sea data. The results show the relevance of the DPC technique to providing synthetic aperture time (phase) compensation even for navigation errors of several tens of wavelengths.

8. VERNIER PROCESSING

The towing speed of synthetic aperture sonar (or any high-resolution sidescan sonar) is limited by its resolution (equation 2.9). Increasing the resolution will thus lead to decreasing the coverage rate or to navigating at lower speed where the tow fish may become less stable. Increasing the coverage rate to resolution ratio is possible thanks to the use of Vernier arrays [24]. By using several hydrophones, Vernier arrays allow either an improvement in towing speed (for a given resolution) or an improvement in resolution beyond the conventional half receive array length, without reducing the azimuth sampling interval.

The principle of Vernier array is presented in figure 2.21:
- Figure 2.21-a shows a standard transmitter/receiver (T/R) configuration for three consecutive transmissions. The synthetic array is correctly sampled (as required by equation 2.26): the spatial sampling is equal to half the receiving array length. By considering a stop-and-go configuration, the history of the phase centres of the array is shown in the lower part of the diagram (each phase centre being in the middle of the T/R array).
- The same system is shown in 2.21-b where the towing speed is twice the limit given by equation (2.26); the bottom image is undersampled.
- .In order to retrieve the sampling requirements with the same high towing speed, it is possible to add a receiver (R) (figure 2.22-c) which is used to recover the samples missing in the second scheme. In this latter case, the phase centre of the additional receiving array can be considered to be halfway between the centres of the T/R and the R parts. The synthetic array is now correctly sampled since the increase in the tow speed is compensated by the increase in the length of the receiver section. It is important to note that because the transmitter length is unchanged, the synthetic array resolution remains the same. Thus, the increase in tow speed does not deteriorate the azimuthal resolution.

The number of additional arrays will mainly be limited by the size of the tow fish and the computation complexity. Vernier array will lead to the improvement in resolution for synthetic aperture sonar without reducing the area coverage rate.

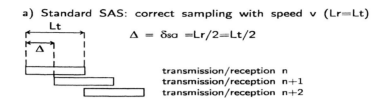
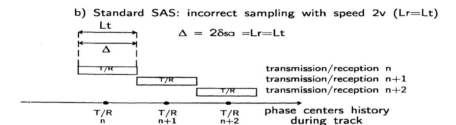
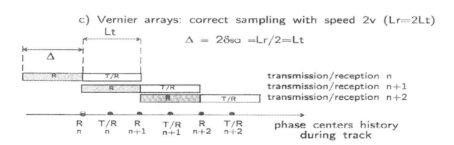

Figure 2.21: Principle of Vernier arrays.

9. THE SAMI PROJECT

9.1 Introduction

The objective of the SAMI project -Synthetic Aperture Mapping and Imaging- was to develop and to test at sea a wideband synthetic aperture sonar prototype capable of providing high resolution seafloor images together with bathymetry maps.

The project has been carried out under the MAST2 (MArine Science and Technology) R&D programme funded by the European Commission [38, 39, 40, 41]. Over a period of three years (September 1993-August 1996) this project gathered five European partners working in close collaboration:
- two university laboratories: CPE Lyon, LASSSO (France) and University of Newcastle upon Tyne, DEEE/Underwater Acoustics Group (UK),
- two research centres: IFREMER (France) and Danish Maritime Institute (Denmark) and
- one industrial partner: Reson A/S (Denmark).

The complementary nature of the partners' specialities and their close collaboration has allowed the ambitious objectives of the project to be achieved. This has included the validation at sea of new concepts of imaging using a special purpose prototype with real-time -on board- processing fully developed within the project.

The project was divided into several tasks. As described in figure 2.23, they range from fundamental research to image interpretation at sea, and include system analysis, development of a simulator, design of signal processing algorithms, tank experiments, construction of a sonar prototype, construction of a real-time processing system, system testing, sea trials and sea data interpretation. In parallel with a software simulator, a tank simulator has been developed that was used for the validation of both simulation and processing algorithms.

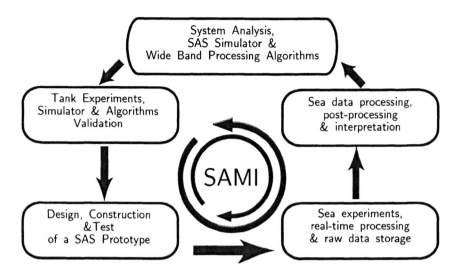

Figure 2.23: Methodology of the SAMI project.

9.2 Validation of algorithms

This section will present several tank experiments in order to illustrate and validate the principles described in the previous sections. All the tank experiments used for illustration (in this paragraph and in the following ones) have been conducted (unless specified) in the following conditions:
- transmitter: 20 mm by 2 mm array,
- receivers: 3 identical 2 cm by 2 mm arrays,
- central frequency: 800 kHz,
- bandwidth: 530 kHz,
- signal: two periods tone burst .
- physical array: 6 mm to 20 mm,
- synthetic aperture length: 100 to 400 mm,

A scale of about a hundred (with respect to the sea trials) was chosen. A computer controlled system was set up and used for:
- signal generation
- trajectory generation and control
- echo digitising and storage.

A 6 degrees of freedom mechanical system (3 rotations plus 3 translations) was constructed for simulating any sonar trajectory. The whole system was fully controlled by the computer and any pre-computed trajectory could have been used.

9.3 Free field experiments

The first series of experiments were conducted using free field targets (spheres and cylinders). Figure 2.23 shows an example of synthetic aperture sonar when the target is omnidirectional (spherical shell). The comparison of raw and processed image can first show an improvement in both azimuthal resolution and signal to noise ratio. The first important echo component is the specular echo. The other "spurious" echo components are due to surface acoustic waves travelling around the target. They are situated at the same delay for every sonar position and their velocity is close enough to the sound velocity in the water so they still posses high amplitude after focusing. Although these spurious echoes may blur the sonar image, they could be of high relevance in a classification phase for the recognition of man-made objects.

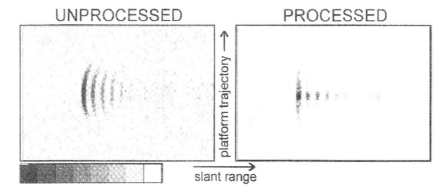

Figure 2.23: Synthetic aperture imaging of a spherical shell (diameter: 15 mm, thickness: 0.45 mm, brass shell). Scale: horizontal [60-65 cm], vertical: [0-12 cm], 4 dB /grey level, maximum in black.

Figures 2.24 and 2.25 show similar results for an extended target (with narrow retransmission beam-pattern): cylindrical shell with spherical endcaps. They show two typical cases for two target attitudes (angle between the target axis and the sonar platform trajectory):

- in the first case (figure 2.24), the main beam of the target intersects the transducer beam; an important echo is obtained in the backscattering mode which does not correspond to the minimum range. After synthetic aperture focusing, the target is located in the correct position and its tilt angle corresponds to 7° (real value).

- in the second case (figure 2.25) the main beam of the target does not intersect the transducer beam: only the echoes of both ends can be seen (with a lower signal to noise ratio). After synthetic aperture, two bright spots corresponding to both ends can be clearly separated but no echo is reflected by the continuous cylindrical part of the target.

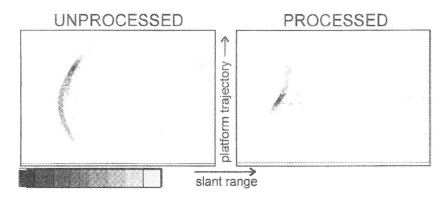

Figure 2.24: Synthetic aperture imaging of an extended (directional) target; attitude: 7°.
Scale: horizontal [103-108 cm], vertical [0-60 cm], 4dB/grey level, maximum in black.

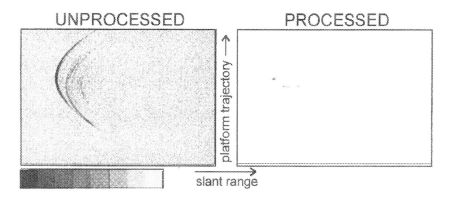

Figure 2.25: Synthetic aperture imaging of an extended (directional) target; attitude: 14°.
Scale: horizontal [107-112 cm], vertical [0-80 cm], 4dB/grey level, maximum in black.

9.4 Target lying on the seabed

The second part of the experimental data concerns target lying on a rough seabed (figures 2.26 and 2.27). For an elevation angle of 25°, the same target attitudes are shown (7° and 14°) for a truncated cylinder (brass shell, diameter = height = 36 mm).

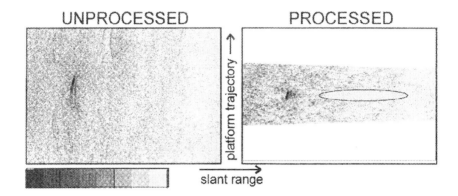

Figure 2.26: Synthetic aperture imaging of a cylindrical shell on a rough bottom; attitude: 7°
  Scale: horizontal [100-119 cm], vertical: [0-40 cm], 5 dB /grey level, maximum in black.

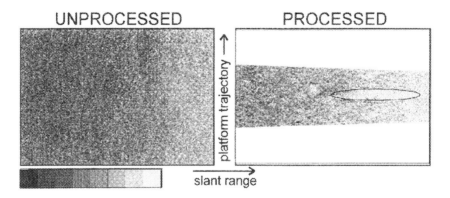

Figure 2.27: Synthetic aperture imaging of a cylindrical shell on a rough bottom; attitude: 14°.
  Scale: horizontal [100-119 cm.], vertical: [0-40 cm.], 5 dB /grey level, maximum in black.

As far as the target reconstruction is concerned, the same observation can be made. In the second case (figure 2.27), the echoes of the cylinder ends are so small that they cannot be discriminated against the bottom reverberation. In both cases, the increase of the azimuthal resolution leads to an improvement of the contrast between the bottom reverberation and the target shadow. In both cases, the target shadow that cannot be distinguished before synthetic aperture processing can clearly be seen after processing even when the target echo cannot be detected.

The shadow shape is ellipsoidal but the shadow area is "corrupted" by some bottom reverberation. This is due to the fact that, for synthetic aperture sonar, the shadow geometry is not the same for every platform position [37]. Indeed, for an elementary scatterer on the bottom, we may be faced to 3 cases:
- the scatterer is never insonified,
- the scatterer is always insonified or
- the scatterer is insonified only for given platform positions.

The last case is only found in the synthetic aperture case and explains the shadow area "corruption" by spurious echoes. For SAS sonar, the shadow structures will thus be different from the one obtained by conventional sidescan sonar possessing the same resolution.

These figures clearly illustrate the interest of synthetic aperture sonar for enhancing the shadow/reverberation contrast. Other tank experiments and simulations have shown that the enhancement of reverberation to shadow contrast is quite proportional the improvement of the azimuthal resolution [37].

9.5 Mapping

The next experimental data concerns the validation of mapping algorithms. A stepped and rough bottom mock up (including a continuous slope) has been constructed [11]. Two transducers were used in parallel (top and bottom). Synthetic aperture was applied to both outputs. Figures 2.28 show both images where high scattering by the steps' roughness can be pointed out (on the left in both figures).

Scattering from the slope roughness can also be observed (at lower level) on the right. The cross-correlation between both high-resolution images is used to build up the high-resolution maps. Figure 2.29 shows the average structure of the stepped mock up (left) together with the reconstructed map of the actual rough one. The data has been used for optimising the integration window cell and show clearly the ability of an optimised system to provide high resolution maps.

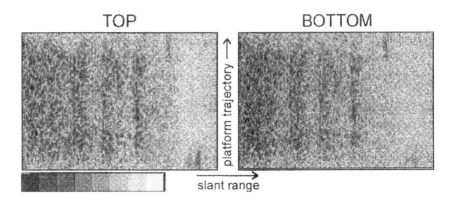

Figure 2.28: Stereo sonar images for bathymetry correlation.
Scale: horizontal [0.6-0.98 m], vertical [0-1 m], 5dB /grey level.

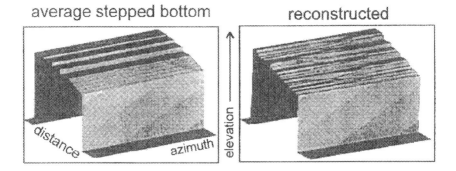

Figure 2.29: Stepped bottom mapping.
Scale: horizontal [0.6-0.98 m], vertical [0-1m], elevation [0-1m].

9.6 Autofocusing

The last tank experiments concern the validation of autofocusing algorithms. The bright spot algorithm was validated for the same spherical shell with a disturbed trajectory. Figures 2.30 show from top to bottom, both raw and processed images for a perfect trajectory (a and b), for a disturbed trajectory (c and d) and for fluctuation corrected trajectory (e and f). Cross-sections at the target position are also displayed. One can first note the high image degradation due to the high frequency trajectory disturbance. One can also point out the high degree of resemblance between the synthetic image computed without disturbances and the one computed after autofocusing algorithm. This resemblance can be reinforced by comparing the image cross-section obtained at the target position for both cases (b and j).

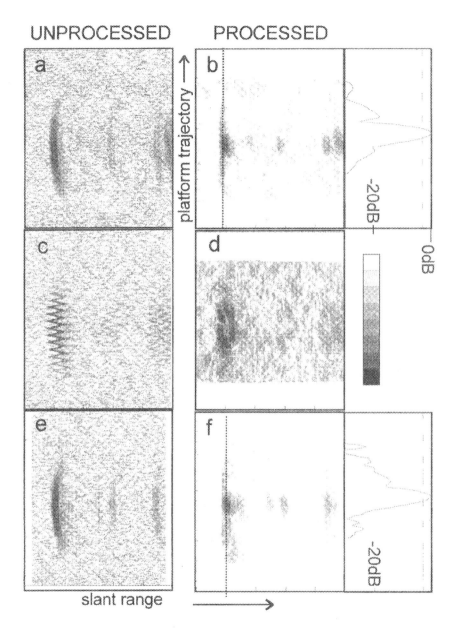

Figure 2.30: Tank validation of autofocusing algorithm on a spherical shell.
Scale: horizontal [0.6-0.98 m], vertical [0-1 m], 5dB /grey level, maximum in black.

9.7 Sea trials: the SAMI prototype

The project has led to the construction of a prototype system that was tested at sea on well-known areas in the Mediterranean for comparison with existing systems. These areas included several types of seabeds (rock, gravel, sand), variable depth areas (from 100 to 2500 m), and several objects (from wrecks to geological structures: rocks fields, sub-marine canyons and mountains, etc.), The towfish was stable enough such that synthetic aperture processing could be achieved without any navigation correction.

The SAMI prototype was divided into two main parts:
- the bottom unit included in the towfish and
- the surface unit used for on board data retrieval, processing, display and storage.

The tow-fish is a neutrally buoyant vehicle comprising:
- the sonar arrays,
- the associated electronics and
- the bottom navigation unit.

It is towed at low speed relatively near the surface (immersion less than 300 m) by an electro-optical cable. The tow-fish is equipped with a pitch and roll adjustment (trim control) operated from the surface using operator feedback on the navigation data.

The array configuration, comprising 4 sub-arrays, is shown in figure 2.31
- (A+B) or (C+D) for standard imaging: 1 m resolution
- (A) and (B) or (C) and (D) for Vernier mode: 0.5 m resolution
- (A+B) and (C+D) for standard imaging with bathymetry.

A surface controller allows the parameters of the transmitter/receivers to be adjusted (including power level, TVG[8] laws, transmitted waveform, receiver gain, ping rate, etc.).

For each channel of sonar data, the real-time hardware comprises a matched filter (time compression) followed by a synthetic aperture processor (azimuth compression). Both of which operate in the time domain.

The matched filtering makes use of a digital signal processor (Motorola DSP 56001) associated to FIR[10] controllers (Motorola DSP 56200), 64 of these controllers were used in order to ensure the following performance: 8192-tap filter which is capable of processing, in real time, chirps with a duration of 100 ms sampled at 80 kHz.

The synthetic aperture machine is a "Transputer" based custom computer capable of forming a synthetic array of 75 metres in real-time (for a swath of approximately 750 m). The heart of this hardware consists of a pipeline of 16 Transputers that performs the focusing operation on selected range intervals [1, 26, 31]. The load balancing is achieved by splitting the data depending on the length of the synthetic aperture at a given range that, in turn, determines the processing requirement at that range.

9.8 Navigation measurements

A bottom navigation unit installed in the tow-fish provides real-time raw navigation data to the surface unit. This navigation system is based mainly around an Inertial Measurement Unit that gives the three components of the acceleration with respect to the body axes and the three components of the absolute rotation rate. These 6 elementary measurements are combined to provide earth referenced attitude and velocity information. Furthermore, a fiuxgate heading sensor, a quartz pressure sensor, and two electromagnetic velocity probes also provide information to the surface navigation unit. This latter processing unit combines all these measurements with those coming from the vessel navigation unit (dGPS[11] positions and velocities, heave, roll, pitch, heading, velocity of the vessel). It provides navigation measurement at both large scale (for map co-ordinates, long term accuracy, i.e. a few meters) and fine scale (for synthetic aperture processing, short-term accuracy, i.e. a few centimetres). Although the best affordable navigation systems were used, the experience showed that the resulting accuracy was not sufficient for synthetic aperture. Fortunately, the fish was stable enough so that synthetic aperture processing could be done with no motion compensation for the main part of the survey.

The characteristics of the SAMI prototype are summarised in following table.

TRANSMITTED SIGNAL	
central frequency	18.2 kHz
frequency range	5-10 kHz (at -6 dB)
transmitted level	218 dB ref. 1/µPa. at 1 m.
Linear Period Modulated (LPM) chirps duration	from 10 to 100 ms
ping repetition period	from 0.340 s to 6.8 s
TVG laws	$30.\log(r) + 1.2r$ $20.\log(r) + 1.2r$

GEOMETRY	
tow fish length	4 m
tow fish immersion	< 250 m.
towing speed	3 knots typ.
operation range	100 to 2500 m (typ. 750 m)
transmitter length	1 m
receiver length	2 m
transmitter/receiver height	0.26 m
separation between rows	0.26 m (centre to centre)
number of receiving channels	4

REAL-TIME PROCESSING	
2 sonar channels and 1 map or imaging + motion compensation	
swath in real time	750 m/channel
theoretical resolution	1 metre (at all ranges)
coverage rate	2.7 km²/hour

The block diagram of the SAMI prototype is shown in figure 4.31.

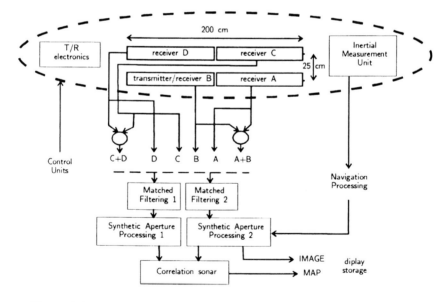

Figure 2.31: Description of the SAMI prototype.

9.9 Validation at sea

Standard reflectors. Before imaging and mapping large areas, the system was tested on standard reflectors placed, during the experiment, on the sea floor in order to quantify the resolution. These targets were quasi-omnidirectional iron reflectors (beamwidth of about one radian at 8 kHz), with high target strength and a relatively small size. Two types of configurations were used, the first included 4 targets fixed on a rigid wooden frame (2 m x 1 m), and the second included 3 separated targets.

Figure 2.32 shows the image of such a target before processing (left) and after processing (right). On the raw images, one can note both the extension of the echoes in azimuth (more than 50 meters at -15 dB), and the hyperbolic range migration. The raw image point spread function is given by the aperture of the physical array, corresponding to a 0.1 radian beamwidth, at the given slant range. Thus, the -6 dB resolution is about 34 meters for 340 meters range. After processing the synthetic array using the average velocity of the ship, the along-track resolution is improved in the focused image. Nevertheless, some ghosts are still visible all around the target. These features are characteristic of a high frequency disturbance of the sonar trajectory with respect to the synthetic array length [15].

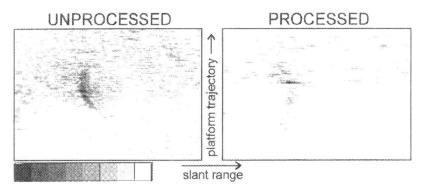

Figure 2.32: Validation of the SAMI system at sea on standard targets.
Scale: horizontal [330-350 m], vertical [100 m], 5dB /grey level, maximum in black.

Autofocusing. The auto-focusing algorithm previously described (the bright spot method) was used to improve the synthetic aperture point spread function. Figures 2.33 shows the received echoes together with the best-fit hyperbola given by the auto-focusing method. After autofocusing, the target images offer a better contrast. It can also be noted that the ghost echoes have disappeared. The gain in resolution can be measured by examining the along-track cross section at the target position as shown in figure 2.33.

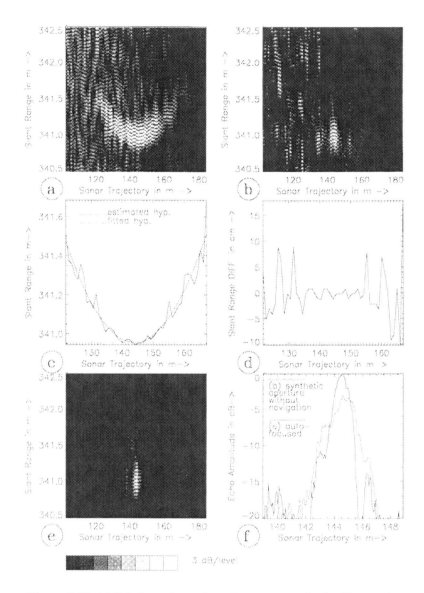

Figure 2.33: Validation of autofocusing on a standard reflector at sea.
Scale: horizontal [340.5-342.5 m], vertical: [110-180 m], 3 dB /grey level, maximum in white.

From the cross-section, one can compute the resolution gain:
- theoretical resolution: 1 metre,
- resolution without auto-focusing: 1.6 ± 0.1 m,
- resolution with autofocusing: 1.1 m ± 0.1 m.

9.10 Some examples

In this section, several images of extended areas are shown that were obtained with the SAMI prototype in the Nice area (France). In figure 2.34, the left image shows the raw data and the right image the processed one. It is important to note that given the scales used (several hundred meters on each direction) the focusing areas (such as the one shown in figures 2.33) correspond to a single pixel in these images. Several sediment features (ridges in the upper and lower part of the images) that could not be seen on the left images are clearly resolved in the right one. Even in the riverbed (image centre) where very small scatterers are encountered, some additional features can still be observed.

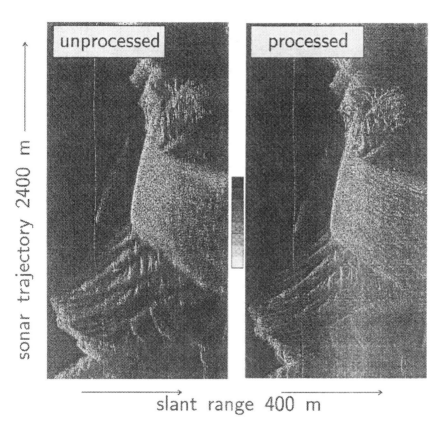

Figure 2.34: Example of sonar image, Nice area.
Scale: horizontal [400 m], vertical: [2400 m], 3 dB /grey level, maximum in white.

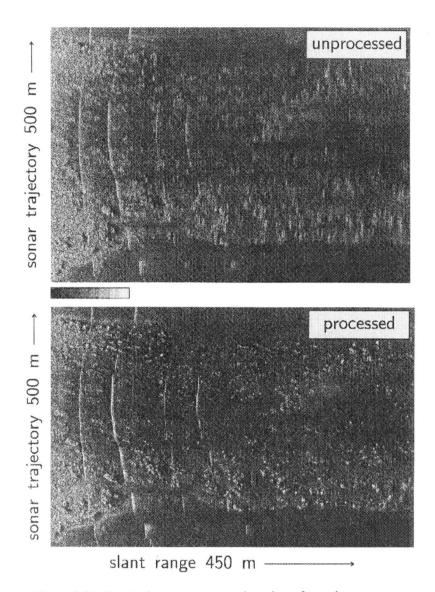

Figure 2.35: Synthetic aperture sonar imaging of a rocky area.
Scale: horizontal [450 m], vertical: [500 m], 3 dB /grey level, maximum in white.

Figure 2.35 is a typical illustration of the interest of synthetic aperture processing and of the importance of dynamic focusing for image interpretation. The top figure represents a row sidescan image in a highly rocky area. In this image, the close rocks are clearly distinguished and their shadow can be clearly identified. As, for conventional sidescan sonar, the resolution is an angular one, the azimuthal resolution will deteriorate with range: the far rocks cannot be resolved and their shadow is blurred. The resolution degradation with range is clear on the top image.

The bottom image represents the result of synthetic aperture processing. All the results presented have been obtained, in real-time, on-board the ship. The azimuthal resolution being constant with range, a rock image does not vary with range. This is clear in the bottom figure. The interest of synthetic aperture increases with range the processing gain (in resolution) ranges from 1 (at 10 meters) to 45 (at 450 meters).

The shadow of far rocks of metric size at 450 metres range can still be clearly distinguished although the sonar is operated in a low frequency range (6-12 kHz). This observation confirms that the image quality is more resolution than frequency dependent.

It is important to note that, for all the images displayed, only synthetic aperture processing has been applied. No additional conventional sonar image processing techniques such as geometry correction, contrast enhancement...was used. This was done, on purpose, to allow objective image comparison and/or measurement on image cross-sections. Additional post-processing can of course be applied to synthetic aperture sonar images (as well as standard ones) in order to improve the image quality.

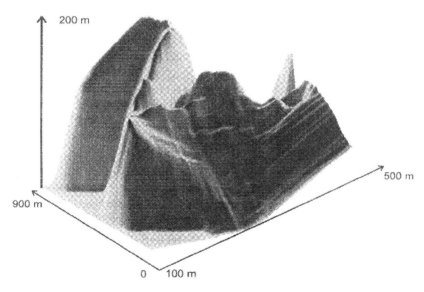

Figure 2.36: Map of the Nice area (SAS +cross-correlation).
Scales: [100-500 m], [0-900 m], elevation: 200 m.

Figure 2.36 shows a map obtained in the Nice area (France) using two receivers, applying synthetic aperture to each receiver output and correlating both synthetic outputs (figure 2.31) .The map is represented in 3D. The volume resolution (voxel) is about a cubic meter (1m by 1m by 1m). The bottom structure (mountains, canyons...) can clearly be seen. The gain in resolution is comparable to previous case.

10. CONCLUSION

This chapter illustrates the interest of using wideband synthetic aperture sonar for bottom imaging and mapping. The processing provides images with a distance resolution that is independent from both the target range and the central operating frequency.

The price to pay can be summarised into two points: the processing complexity and the platform stability.

As far as the processing complexity is concerned, the use of a wide frequency range and an appropriate signal design can lead to a considerable decrease of the processing complexity. Nevertheless, it may lead to signal oversampling requirements that may make the baseband processing as complex as the processing around the carrier (RF). Nevertheless, the increase of the computation power and the possibility of algorithm parallelisation can overcome this difficulty.

The Achilles' heel of synthetic aperture remains the navigation of the sonar platform. Even when a very stable platform is used, additional navigation sensors are required. Standard sensors, such as inertial navigation units, may provide a first (rough) information on the trajectory but are not accurate enough for full correction, in most cases. Additional informations on the trajectory -based on autofocusing techniques- are needed in order to provide the final trajectory corrections. Several methods are proposed and validated in the literature but research work is still undertaken to improve the accuracy of such methods.

Providing high-resolution images (or maps) is faced to the spatial sampling requirement problem. This requirement may lead to the reduction of the towing speed (and thus the coverage rate). The Vernier array technique has been used successfully to overcome this problem.

As the synthetic images are frequency independent, they may be used for mapping the same area with the same resolution in different frequency ranges. If the frequency range is wide enough (or covers several sub-bands), a wide band spectral characterisation of the bottom is possible. This may open new research trends for bottom characterisation that were not possible with conventional sidescan sonar where the resolution varies with frequency.

11. ACKNOWLEDGEMENTS

The work was achieved, at CPE Lyon, in the framework of the SAMI project, MAST2 (MArine Science and Technology) program funded by the European Commission. The authors thank all the partners who have participated in the project: University of Newcastle upon Tyne -DEEE- Underwater Acoustics Group (UK), IFREMER (France), Danish Maritime Institute (Denmark) and Reson A/S (Denmark).

12. NOTES

1. CTFM: Continuous Transmission Frequency Modulation.
2. SAS: Synthetic aperture Sonar.
3. GESMA: Groupe d'Études Sous-Marines de l'Atlantique.
2. ACID: ACoustical Imaging Development (EC funded project).
5. SAMI: Synthetic Aperture Mapping and Imaging (EC funded project).
6. AUV: Autonomous Underwater Vehicle.
7. ROV: Remote Operated Vehicle.
8. TVG: Time Varying Gain.
9. DSP: Digital Signal Processor.
10. FIR: Finite Impulse Response (filter).
11. dGPS: differential GPS (Global Positioning System).

13. REFERENCES

[1] Adams, A., Lawlor, M., Riyait, V., Hinton, O., and Sharif, B. (1996). A real-time synthetic aperture sonar system. IEE Proceedings on Radar, Sonar and Navigation, 143(3): 169-176. Special issue: Recent Advances in Sonar and Its Applications in the Ocean.

[2] Altes, R. (1976). Sonar for generalized target description and its similarity to animal echolocation systems. J. Acoust. Soc. Am., 59 (1): 97-105.

[3] Altes, R. and Reese, W. (1975). Doppler tolerant classification of distributed targets -a bionic sonar. 11(5): 708-722.

[4] Bilge, H., Karaman, M., and O'Donnel, M. (1996). Motion estimation using common spatial frequencies in synthetic aperture imaging. In IEEE Int. Ultrason. Symp, pp. 1551-1554, San Antonio, Texas (USA).

[5] Blacknell, D. and Quegan, S. (1990). Motion compensation of airborne synthetic aperture radar using autofocus. J. Geophys. of Res., 7(3):168-182.
[6] Bouhier, M. and Zakharia, M. (1990). ACID: A MAST project on ACoustical Imaging Development. Oceanology International 90, Brighton (United Kingdom).
[7] Châtillon, J. (1994). Application de la synthèse d'ouverture en sonar actif. PhD thesis, INSA de Lyon (France).
[8] Châtillon, J., Adams, A., Lawlor, M., and Zakharia, M. (1999). SAMI: A low frequency prototype for mapping and imaging of the seabed by means of synthetic aperture. IEEE J. on Ocean. Eng., 24(1): 4-15.
[9] Châtillon, J., Bouhier, M., and Zakharia, M. (1992). Synthetic aperture sonar for seabed imaging: Relative merits of narrow band and wideband approaches. IEEE J. on Ocean. Eng. 1 17(1): 95-105.
[10] Châtillon, J. and Zakharia, M. (1996). Self-focusing of synthetic aperture sonar in case of bottom reverberation. In Papadakis J., Editor, Third European Conference on Underwater Acoustics, pp. 433-438, Heraklio, Crete, Greece. European Commission, Brussels, Belgium.
[11] Châtillon, J. and Zakharia, M. (1996). Validation of bathymetry algorithms using wideband synthetic aperture techniques by means of tank experiments. In Papadakis, J., editor, Third European Conference on Underwater Acoustics, pp. 427-431, Heraklio, Crete, Greece. European Commission, Brussels, Belgium.
[12] Châtillon, J., Zakharia, M., and Bouhier, M. (1991). Quantification of the quality of images obtained by synthetic aperture sonar. Proc. of the IOA, 13(9), pp. 147-152.
[13] Châtillon, J., Zakharia, M., and Bouhier, M. (1991). Synthèse d'ouverture en acoustique sous-marine: influence de l'effet Doppler différentiel. In proc. of Treizième Colloque du Groupe de Recherche et d'Étude de Traitement du Signal-GRETSI-, pp. 601-604, Juan-les-Pins (France).
[14] Châtillon J., Zakharia, M., and Bouhier, M. (1992). Navigation inaccuracies in synthetic aperture sonar: simulations and experiments. In Undersea Defence Technology '92, pp. 553-557, London (United Kingdom). Microwave Exhibitions and Publishers Ltd.
[15] Châtillon, J., Zakharia, M.) and Bouhier, M. (1994). Self-focusing of synthetic aperture sonar: Validation from sea data. In Bjørnø, L., editor, Second European Conference on Underwater Acoustics, pp. 727-731, Lyngby (Denmark). European Commission, Brussels, (Belgium).
[16] Curlander, J. and McDonough, R. (1991). Synthetic Aperture Radar. John Wiley, New-York. 647 p.
[17] Cutrona, L. (1975). Comparison of sonar system performance achievable using synthetic aperture techniques with the performance

achievable by more convention al means. J. Acoust. Soc. Am., 58(2): 336-348.
[18] de Heering, P.) Simmer, K., Ochieng-Ogolla, E., and Wasiljeff, A. (1994). A deconvolution algorithm for broadband synthetic aperture data processing. IEEE J. on Ocean. Eng., 19(1):73-83.
[19] Gough, P. and Hayes, M. (1989). Tests results using a prototype synthetic aperture sonar. J. Acoust. Soc. Am., 6(6):2328-2333.
[20] Griffiths, J. and Gida, A. (1984). Use of a BBC microcomputer for synthetic aperture measurements. Proc. Of the IOA 6(6):122-128.
[21] Guyonic, S. (1994). Experiments of a sonar with a synthetic aperture array moving on a rail. In Oceans '94 Conf Record, pp. 571-576, Brest (France). MTS and IEEE publishers.
[22] Huxtable, B. and Geyer, E. (1993). Motion compensation feasibility for high-resolution synthetic aperture sonar. In Oceans 93 Conf Record, pp. 1.125-1-137. MTS and IEEE publishers.
[23] Johnson, K., Hayes, M., and Gough, P. (1995). A method for estimating the sub-wavelength sway of a sonar towfish. IEEE J. on Ocean. Eng., 20(4):258-267.
[24] Kock, W. (1972). Extending the maximum range of synthetic aperture (hologram) systems. Proc. IEEE (Lett.), 60(11):1459:1460.
[25] Lawlor, M., Adams, A., Hinton, 0., Riyait, V., and Sharif, B. (1994). Methods for increasing the azimuth resolution and mapping rate of a synthetic aperture sonar. In Oceans 94 Conf. Record, pp. 565-570, Brest (France). MTS and IEEE publishers.
[26] Lawlor, M., Hinton, 0., Adams, A., and Sharif, B. (1992). Design of a real-time parallel processing system for synthetic aperture sonar processing. In Undersea Defence Technology 92, pp. 275-280, London (United Kingdom). Microwave Exhibitions and Publishers Ltd.
[27] Loggins, C., Christoff, J., and Pipkin, E. (1982). Results from rail synthetic aperture experiments. J. Acoust. Soc. Am., 71:85. suppl. 1.
[28] Mamode, M. (1981). Estimation optimale de la date d'arrivée d'un écho perturbé par l'effet Doppler. Synthèse de signaux large bande tolérants. PhD thesis, INP Grenoble (France).
[29] Raven, R. (1981). Electronic stabilization for displaced phase centres systems. Technical report. U.S. patent 4244036.
[30] Riyait, V., Lawlor, M., Adams, A., Hinton, 0., and Sharif, B. (1994). Comparison of the mapping resolution of the acid synthetic aperture sonar with existing sidescan sonar systems. In Oceans '94 Conf. Record, pp. 559-564, Brest (France). MTS and IEEE publishers.
[31] Riyait, V. Lawlor M., Adams A., Hinton O., and Sharif, B. (1995). Real-time synthetic aperture sonar imaging using a parallel architecture. IEEE Trans. on Imag. Proc., 4(7): 1010-1019.

[32] Rolt, K., Milgram, J., and Schmidt, H. (1994). Broadband undersampled synthetic aperture arrays: targets stay sharp, aliases smear. Cambridge MA (USA). Acoustical Society of America 127th meeting. Abstract number: 2UW18.
[33] Rolt, K. and Schmidt, H. (1992). Azimuthal ambiguities in synthetic aperture sonar imagery and synthetic aperture radar imagery. IEEE J. on Ocean. Eng., 17(1):73-79.
[34] Sato, T. and Ikeda 0. (1977). Sequential synthetic aperture sonar system -a prototype of a synthetic aperture system. IEEE Trans. on Son. and Ultrason., SU-24(4):253-259.
[35] Sheriff, R. (1992). Synthetic aperture beamforming with automatic phase compensation for high frequency sonars. In Symposium on Autonomous Underwater Vehicle Technology, pp. 236-245.
[36] Tonard, v. and Brussieux, M. (1997). Towards development of autofocusing schemes for phase compensation of synthetic aperture sonars. In Oceans 91 Conf Record, pp. 803-808, Halifax (Canada). MTS and IEEE publishers.
[37] Tonard, V. and Châtillon, J. (1997). Acoustical imaging of extended targets by means of synthetic aperture sonar. Acustica united with Acta Acustica, 83(6): 992-997.
[38] Zakharia, M. and Châtillon, J. (1994). A low frequency wideband synthetic aperture sonar prototype. Cambridge MA (USA). Acoustical Society of America 127th. meeting. Abstract number: 4aUW17.
[39] Zakharia, M. and Châtillon, J. (1994). Wideband synthetic aperture sonar for bottom imaging. Theoretical aspects and experimental evaluation. Oceanology International 94, Brighton (United Kingdom). vol. 3.
[40] Zakharia, M. and Châtillon, J. (1995). Synthetic Aperture Mapping and Imaging- SAMI. In Second MAST Days and EUROMAR market, pp. 1161-1171, Sorrento (Italy).
[41] Zakharia, M. and Châtillon, J. (1997). SAMI: a low frequency wideband prototype for synthetic aperture mapping and imaging. Penn State PE (USA). Acoustical Society of America 133rd meeting. Abstract number; 4aUW3.
[42] Zakharia, M. and Guigal, A. (1991). Étude et description de signaux tolérants à l'effet Doppler variable. In 13ème Colloque GRETSI, pp. 597-600, Juan les Pins (FR).

3

Integrated Programmable Underwater Acoustic Biotelemetry System

R.S. H. Istepanian
E-Med Systems and Health Engineering Group
Department of Electronic & Computer Engineering,
Brunel University, Uxbridge, Middlesex, UB8 3PH, UK
E-mail: Robert.Istepanian@brunel.ac.uk

1. INTRODUCTION

Biomedical telemetry (Biotelemetry) is a special facet of bioinstrumentation, which provides a means for transmitting physiological or biological information from one site to another for data collection. Technically, it refers to systems, which require no mechanical connection. The actual or encoded parameters are usually transmitted via acoustic or radio waves, although light waves have also been used. Biotelemetry studies in the last three decades have permitted many areas of physiological and behavioural monitoring in diverse conditions, both for humans and animals, without the encumbrance and restriction of wires connecting the transmitter and receiver. The most widespread use of biotelemetry is the monitoring of biological information from animals and man. The importance of biotelemetry to basic biological, environmental, and medical research cannot be overstated. For example, the utility to provide real time physiological telemetry monitoring in the hospitals has become widely recognised since the early 1970's.

Biotelemetry is also used as an aid to understanding and identifying the natural causes that are linked to habitat conditions of wild animals, which in turn alter their behaviour, and how such conditions affect their mortality rates. It can also provide a means to study and predict the effects of environmental changes such as thermal and chemical pollution and other geophysical changes. One of the important areas in which biotelemetry is useful is in human underwater physiological monitoring. It is of prime importance to studies in physiology and exercise, physical stress and many other effects, and is particularly applicable to monitoring swimmers and divers.

There are many other useful applications for such systems; for example in a sporting context, athletes may want to monitor important physiological responses under quasi-competitive conditions. In the case of divers, acoustic biotelemetry enables the monitoring of different evoked potentials and physiological parameters from freely moving subjects performing different underwater tasks. However, once a decision has been made to adopt acoustic telemetry, there are several considerations that have to be taken into account when designing a system. The main problem, especially in very shallow water environments such as swimming pools, is that of multipath acoustic reverberation. However, recent advances in high speed digital single-chip microcontrollers and digital signal processors (DSPs) have made it possible to overcome the various obstacles that hindered progress in the past. This chapter presents a comprehensive new design methodology for underwater acoustic biotelemetry for engineers and physiologists interested in this important research area. The concepts of designing and implementing programmable multi-channel digital acoustic biotelemetry systems using

DSPs and single-chip micro controller systems for real-time physiological monitoring of free swimming subjects and divers. A system like this was difficult to implement in the past, but recent advances in high-speed digital zsignal processors and single-chip technology have made such an application achievable.
The advantages of this new design approach are that it provides a telemetry link through its unique modular architecture. This design comprises flexible hardware modules and programmable software to accommodate the performance requirements and overcome the fundamental problems associated with realising such systems. It also allows the design specifications to be generalised for future applications.

2. AN OVERVIEW OF UNDERWATER BIOTELEMETRY SYSTEMS

Numerous studies have been carried out to develop systems for telemetering physiological data through air (terrestrial telemetry). Most of the biopotentials such as Electrocardiogram (ECG), Electroencephalogram (EEG), Electromyographic (EMG) signals, and other physiological variables such as temperature, respiration, blood pressure etc. have been telemetered during the last three decades from human and animals [1,2,3]. However, the emphasis in recent years has been on the development of sophisticated designs of implanted and miniaturised multichannel radio biotelemetric systems [4,5]. Such telemetry studies provide invaluable information about animals in the wild. For example it allows comparison of ecological and physiological parameters under restrained and unrestrained conditions for different species.

It also provides similar information about humans under certain environmental conditions that is otherwise unavailable in experimental data [6,7]. This evolutionary approach in aerial radio frequency (RF) biotelemetry systems was not paralleled with a corresponding advance in underwater acoustic biotelemetry and certainly not for monitoring human subjects. This has been due to several constraints, emphasised in the literature, which are mainly due to the problematic nature of the underwater acoustic transmission and reception. The most serious difficulty is to extract the wanted signals from the actual received signals, which is complicated by multipath interference. Thus, the main aim of this review is to concentrate on the underwater biotelemetry systems used so far and to provide a comprehensive study on the different telemetry techniques applied both on humans and aquatic animals, because of the similarities involved in both applications.

2.1 Biotelemetry systems from Aquatic animals

The first pioneering report on the electrophysiological study of free swimming fish was reported in 1938 [8]. But it was only in the early 1960's that considerable effort was expended on behavioural and physiological radiotelemetry studies of aquatic animals, such as fish, porpoises, turtles and sharks [9,10].

Since then , many electrophysiological studies using radiotelemetry techniques on free swimming fish have been reported [11,12,13,14]. The most common modulation method used was a simple Frequency Modulation (FM/FM) scheme, with the subcarrier signals modulated separately, and then combined to modulate the main carrier frequency. However, several disadvantages were discovered with these systems, such as small range, minimum battery life and the stability of VHF circuits when encapsulated. In addition, there is the very high attenuation of electromagnetic radiation in seawater. This confines such systems to very low frequencies, short ranges and fresh water telemetry applications.

Acoustic transmission offers the advantages of little transmission loss and propagation absorption for ranges up to several hundred metres. Such telemetry systems encouraged researchers to investigate the different behaviour, physiology and locomotive activities of fish species, and other marine animals in their natural habitats. Ultrasonic biotelemetry has its beginnings in 1956 [15] and is sometimes referred to as underwater or marine biotelemetry. Ultrasonic transmitters ('Pingers') have been used in a variety of applications for studying the migration, ecology and physiology of aquatic animals such as fish, mammals, reptiles and invertebrates.

Some of these studies provided a wide range of data on free ranging-animals, such as behavioural information (e.g. habitat utilisation, energy expenditures), water temperatures, depth, swimming speed and thermal preferences [16,17,18,19,20,21]. Physiological research is another aspect of marine ultrasonic biotelemetry, which can provide the critical information on how animals are actually behaving and utilising their natural habitat. Small ultrasonic tags have been used to measure heart rate activity in large vertebrate's [22] and wild brown trout [23]; others have been used to measure changes in the electrical impedance between two electrodes, and transmit activity of a moving part, e.g. heart rate or scaphogathite beat [24]. Electromyograms (EMG) of fish and tail-beat frequency are other bioelectrical activity patterns that are monitored by employing ultrasonic telemetry systems [25,26]. Data obtained from these physiological parameters were correlated with other environmental and field activity factors. However, to date few reported studies exist about the marine acoustic biotelemetry from several sensors.

The behavioural and environmental data of sharks at sea acquired from multiplexed channels measuring swimming speed, depth, compass heading, ambient light and temperature has been reported to have been acoustically telemetered [28]. The behavioural and environmental factors of salmon and codfish in the sea, and also when contained in a tank have also been monitored by standard acoustic telemetry [29,30]. Such research can be important for future fish farming purposes. However, there remain certain limitations on the use of both ultrasonic and radio telemetry systems in the remote sensing of aquatic animals.

Ultrasonic telemetry is well suited for studies in salt water, fresh water with high conductivity and deep water. On the other hand, ultrasonic signals are affected by the presence of macrophytes, algae, raindrops and man-made noise in natural habitat [32]. Also ultrasonic telemetered signals are attenuated by trapped air bubbles, especially in turbulent water and refracted downward by the effect of the thermocline or temperature gradients in warmer waters [33,34]. Radio telemetry, on the other hand, is suited for shallow, low conductivity fresh water, and for turbulent water environments. Provided the low frequency (100 kHz or less) receiving antennas are of small size and do not require contact with water, radiotelemetry is excellent for searching large areas to find highly mobile species (e.g. salmon) [11,14]. Also, radio signal levels are little affected by the presence of vegetation, algae or thermocline. The disadvantages of radio biotelemetry is that it can not be used in salt water unless the subject or the animal swims or surfaces periodically, and the signal levels are attenuated by increasing depth and conductivity[34,35]. Additionally, the radio signals are deflected by metal objects and by terrain, which adds to its range and depth limitations. Furthermore, with depths greater than 5m and conductivity greater than 400 $\mu S/m$, radio telemetry will not work, especially for highly mobile aquatic species. However, in general there exist some common features for both animal biotelemetry transmitters when used in environments for which they are both suited. For example, transmitters of both types could have approximately the same battery life, housing, size, cost and encapsulation.

As a conclusion, the specific physiological, and behavioural information from fish and other aquatic animals requires acoustic telemetry transmitters to be as miniaturised as possible, to last for as long as possible and to provide long range capability. The advanced level of integrated and hybrid technology nowadays can satisfy the requirements of micro-miniaturisation and multichannel automation of marine underwater acoustic biotelemetry. This could bring this branch of biotelemetry in line with advanced aerial biotelemetry. However, there may be more differences than anticipated between fish-tracking technology and that needed for physiological monitoring of divers. The data rates are usually very low in fisheries telemetry, often only a simple 20ms ping once per second for most of the physiological or behavioural data required. Also the multipath problem in the open sea or less reverberant habitats does not represent a real problem for such low rate transmission.

2.2 Biotelemetry systems for swimmers and divers

The first attempts to telemeter and monitor physiological responses of human subjects in relation to their underwater activity, started in the early 1960's with the advent of the space program [1,36]. Since then various biotelemetric methods have been devised for relaying physiological parameters during swimming and diving. The three commonest telemetry methods used were the ultrasonic, electromagnetic and conductive. Considerable effort has been expended on electromagnetic telemetry using very high frequency (VHF) transmitters (up to 100 MHZ) to obtain physiological data from competitive swimming subjects [37,38,39]. However, since in fresh water, the velocity and wavelength of electromagnetic radiation is reduced by about nine times, low attenuation can be easily achieved in swimming pools at frequencies up to at least 1MHz [40]. On the other hand, propagation of electromagnetic radiation in salt or chlorinated water is severely affected by the flow of conduction currents, with further reduction of velocity and a very high attenuation rate [41]. This discourages the use of VHF carrier telemetry even in swimming pool applications [42]. Several low frequency (400 kHz) RF telemetry systems with 'inductive loop' techniques were used for telemetering different physiological parameters in swimming pools [42,23,44]. However, these experiments were carried out under controlled and restricted condition dependent on circular wire loops attached to the subject and wound around the diameter of the swimming pool. Additionally, the range of the low frequency wavelength used was limited and the received signal was affected by interference caused from radio broadcasting channels. An alternative method used for underwater biotelemetry has been the conductive method with a frequency-modulated current as a carrier [46]. This method, also referred to as the 'current return density' method, has been used earlier for the transmission of the gut temperature from dolphins using a 59 kHz carrier frequency [1], and for the electroencephalographic transmission from SCUBA (*self c*ontained *u*nderwater *b*reathing *a*pparatus) divers [47].

This method, however, proved to be hazardous, especially when the electrode contacts necessary for the return current transmission presented electrocution paths if the return current density exceeded a certain level. Additionally, the signal could not propagate in air, i.e. when the subject was not immersed completely. To compromise between the advantages and weaknesses of both approaches, an amphibious ECG telemetry system using a combination of electromagnetic and conductive (return) current transmission in air and water was developed [48]. This system proved viable for transmitting ECG signals from a swimmer in a swimming pool, but it involved a complicated system design in both the transmission and reception circuitry. This limited multichannel implementation using such a technique, where circuit size, weight and power supply is crucial elements in the design. In addition, the system retained the hazardous effect of high return currents into the electrodes attached to the subject, and also involved a cable connection attached to the swimmer that limited the range of transmission. Most of the earliest studies on the physiological and metabolic responses of SCUBA divers were done largely under controlled conditions, where the subjects where tethered to recording cables in hyperbaric chambers and special tanks [49,50.51]. These studies do not necessarily represent the same combination of environmental conditions that face freely swimming divers, and thus reflect the same physiological responses.

Other studies have been carried out with waterproof magnetic tape recorders attached to the diver for recording the cardiac changes and other physiological parameters [52,53]. However, such data can only be examined after the dive, which does not allow observation of the changes in the diver's response in real time. Although ultrasonic transmission is considered the most suitable method for underwater biotelemetry from freely swimming divers for the reasons explained earlier, especially for greater depths and longer distances. The amount of the reported research work carried out since the 1960's in this context are very limited. Acoustic transmission of the ECG from SCUBA divers in the open sea using a direct FM carrier transmission at 55 kHz was reported during the Sea Lab project in the United States [54,55]. Acoustic telemetry has been also used with a group of Hawaiian SCUBA divers equipped with a combination of ECG and respiration transmitters; in order to measure the physiological stress and reflexes they endured while setting fish traps at a depth of 60m [56]. In this method a frequency-modulated carrier of 32.8 kHz was used to relay the electrocardiogram and respiration signals in the open ocean. On the receiver side, a manually tuneable super-heterodyne system with filters was used to separate the mixed high frequency signal at 455 kHz from the carrier signal. Such receivers were largely dependent on a multipath-free underwater medium. A similar approach was adopted to develop a simple multichannel acoustic biotelemetry system [57]. In this work each physiological channel was allocated a different transmission carrier frequency. At the receiver end, the output of the hydrophone was applied to four identical super-heterodyne systems to retrieve the carrier signals. Simple gate logic circuits were used for the sampling between the channels, and each channel was sequentially sampled at a rate of 10 times per second. However, no clear results were reported because the transmission circuitry was prone to timing errors and missing sampled data due to the simple logic design involved. In addition, the channel selectivity of the receiver was manually operated, and unsuitable for use in severe multipath environment for retrieving analogue signals such as ECG signals. It can be seen from this review that although some attempts were carried out for the acoustic physiological transmission and monitoring in open water. However, no real progress has been made in recent years and since the 1990's to design and implement an integrated acoustic multichannel underwater acoustic biotelemetry system for human physiological monitoring that will overcome the main drawbacks of acoustic telemetry especially in shallow underwater channels. To achieve this aim it is now possible to accommodate recent advances in microtechnology and digital signal processing devices to realise a versatile programmable telemetry system for such purposes.

3. UNDERWATER PROGRAMMABLE ACOUSTIC BIOTELEMETRY

In this section we present the main aspects involved in the design and implementation of a digital telemetry system with a reliable detection format and a multipath immune transmission link, particularly for severe multipath environments. It also details the additional limitations imposed for biotelemetry applications. Further, it discusses the various aspects of coding and multiplexing used in radio biotelemetry and their limitations in an underwater environment. The chapter also presents the physiological parameters suitable for underwater biotelemetry and those considered for transmission in the present work. A study of these aspects and the various trade-offs involved in such application has lead to the design philosophy and development of the present biotelemetry system.

3.1 System design considerations and limitations

It is essential to consider the limitations involved in the design and implementation of a particular multichannel acoustic biotelemetry system, as envisaged in the present application. First, it is useful to outline the principal aspects involved in the design of such systems geared for shallow reverberant environments. Secondly, it is worth considering the performance limitations involved in adopting the design principles and transmission formats of a radio biotelemetry system in such an environment. The compromise between the two main design aspects and the general requirements for underwater biotelemetry lead to the final design methodology.

In general, designing an underwater acoustic telemetry system, it is necessary to partition the problem into two main design aspects:

1. Reliable detection format: this includes the range, operating frequency, acoustic power and other related parameters formulated from sonar equations.

2. Multipath immune communication link: this represents the acoustic transmission format required to overcome the multipath problem in shallow underwater environments and the data encoding and modulation techniques that match the data source and allows detection of the wanted signal (direct path) from multipath reverberations.

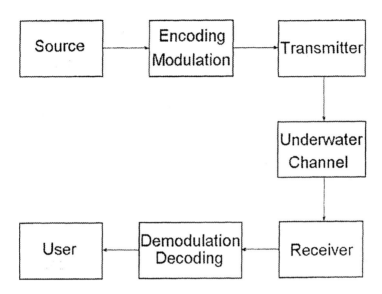

Fig.1   Generalized underwater acoustic communication system

Fig. (1) shows a general representation of a typical underwater acoustic communication system. It consists of the main blocks for an underwater digital data telemetry architecture. In this system, the two design criteria mentioned above are interrelated where the final selection and instrument implementation has to take into consideration all the parameters involved. These will lead to a fundamental trade-off between the various design parameters involved, such as data rate, range, transmission frequency and bandwidth. However, in an underwater biotelemetry design, such limitations are not the only ones that have to be considered. There are additional performance and implementation factors that specify the problem and add further to the complexity of such systems.

The main functional requirements for an underwater multichannel biotelemetry system are as follows:

i) The design and organisation should be flexible enough to accommodate any combination of biopotentials and/or electrically transduced physiological parameters with a minimum of re-engineering effort.

(ii) The ability to acquire as many independent channels as possible.
(iii) Detection of the appropriately timed, activated physiological events.
(iv) Sufficient electronic sophistication and miniaturisation.
(v) Batteries with a life commensurate with the power requirements of the transmitter circuitry.

This section presents a study of all these considerations, together with the different limitations and the other factors, which have lead to the present modular design philosophy.

3.2 Relevant underwater detection and Sonar equations

In order to establish the design of the system to satisfy the first criterion it is necessary to examine the different requirements and constraints involved and to group them according to the overall effects of the environment and the performance of acoustic telemetry.

To do this, calculations can be made using standard sonar equations [34,58,59]:

$$SL + DI_T = NL - DI_R + TL + DT \qquad 1$$

where:
SL is the source level,
DI_T is the transmitter directivity index for a directional source,
TL is the transmission loss between the transmitter and receiver,
NL is the isotropic noise level,
DI_R is the receiver directivity index
DT is the detection threshold (Recognition Differential).

Although the sonar equation (1) is derived for an active sonar, many of the parameters are applicable to passive applications and it can be expressed in telemetry applications to represent the present set-up systems where the signal travels in one direction and the effects of the ambient (rain,ocean surface,shipping noise) for short ranges is not predominant and can be neglected. Therefore Eq.(1) can be modified to represent the present application as:

$$SL = RL + TL \qquad 2$$

where: RL is the received signal level.

In the following the parameters involved in the sonar equation are briefly explained. The source level is a measure, in dB referenced to 1 $\Box Pa$, of the power flux (W/m^2) delivered into the water by a source and is always referred to a standard range (1m) from the acoustic centre of the source.

The source level is related to the acoustic power P_a by:

$$SL = 170.8 + 10\log P_a + DI_T \quad dB \qquad 3$$

For the omnidirectional traducers used here (D1/170, Universal Sonar Ltd., U.K.), the directivity index is zero for both transmission and reception.

The acoustic output power is the product of the electrical Power P_e and the transduction efficiency i.e ($\eta = P_a/P_e$); therefore source level can be written in terms of the electrical power as:

$$SL = 170.8 + 10\log(P_e . \eta) \quad dB \ re \ 1\mu P_a \ at \ 1m \qquad 4$$

where:
$$P_e = V_t^2 / R_{eq} \qquad 5$$

V_t is the r.m.s. voltage applied to the transmitting transducer (projector), R_{eq} is the equivalent resistance of the hydrophone.

This power, conventionally reckoned at 1m from the 'acoustic centre' of the source, is reduced progressively between the projector and the hydrophone by the geometrical spreading and frequency dependent absorption. This is described collectively as transmission loss (TL). For short ranges, the power is reduced by spherical spreading as the reciprocal of the square of the range r, i.e

This can be represented by:

$$TL = 20\log r + \alpha r \qquad 6$$

where α is the absorption coefficient, which is frequency dependent.

Given an absorption coefficient of $\alpha = 0.022$ dB m^{-1} (22 dB km^{-1}), corresponding to a frequency of 70 kHz used here, the transmission loss is thus given by:

$$TL = 20\log r + 0.022r \qquad 7$$

The principal approach in this design criterion is a study of the trade-off between the various requirements and constraints calculated from the above equations leading to the best choice.

The main design parameters that have to be considered in the design of underwater acoustic biotelemetry systems are:
1. Acoustic Propagation in shallow water.
2. Choice of Transmission frequency.
3. Signal level and range calculations.

The details of the first two factors are beyond the scope of this chapter and are given in several text references [34,59,60, 62]. However, we will give a briefly introduce some of the design details of the third parameter for completeness and its direct relevance to this application.

To select the signalling for an acoustic telemetry system appropriate for a reverberant shallow channel, it is important to calculate the expected receiver level to determine the level of noise masking the detected signal. This will also allow the design of the receiver pre-amplifiers and filters. Also, the limitation of a maximum range of operation is essential if reliable telemetry is required. Once the maximum range is determined, the acoustic source level to achieve it may be calculated from values for receiver sensitivity, the required signal-to-noise ratio and transmission loss. The expected receiver level is determined considering the required acoustic output power and the specified range of operation at a given frequency. The design began with the determination of the required transmitter power to achieve a satisfactory signal-to-noise ratio at the receiver for an assumed experimental range of 100m.

For the specified range and the absorption coefficient given at 70 kHz, Eq. (7) yields a transmission loss of :

$$TL = 20 \log(100) + 0.022(100) = 42.2 \text{ dB}$$

For example from the specifications of the omnidirectional transducers used in current system (D1/170, Universal Sonar Ltd., and UK) that each hydrophone has sensitivity R_s at 70 kHz of -201 dB re 1V μPa^{-1}.

The appropriate electrical parameters here are: an electrical power of 0.4 W, and an equivalent hydrophone resistance R_{eq} of 276Ω, and an assumed transduction efficiency of 80% , i.e $\eta = P_{ac}/P_e = 0.8$. Substituting these values in Eqs.(5), (4) and (1), the expected level of the received signal can be calculated as:

$$SL = 170.8 + 10 \log(V^2/R_{eq}) \qquad \text{dB re } 1\mu Pa$$

hence ;

$$20 \log (Vr/R_s) = 20 \log(Vt) + 104.2$$

Substituting for receiver sensitivity (in dB) yields:

$$20 \log(V_r/V_t) = -96.8 \text{ dB}$$

For a designed transmitted voltage of approximately 30 V_{p-p} voltage, Vr should be 0.45 mV_{p-p} at 100m maximum range. Similarly, the expected received voltage at the hydrophone is approximately 60 mV_{p-p} at the 4.5m range suitable for the experimental underwater channel of the test tank platform that will be discussed later. This design example shows the relevant sonar calculations for efficient telemetry for a maximum range of 100m assumed in the current experimental system.

These calculations also show that after a pre-amplification stage and with suitable gain adjustment, the predicted signal level at the hydrophone is easily detectable at the farthest range of the system. However, less gain adjustment is required in the latter case for shorter ranges. This concludes that the restrictive parameters assumed at this example fulfils the design criteria within the operational requirements assumed by the system designer.

3.3 Acoustic communication channel and transmission format

The effects of the multipath in short-range shallow channel result in a taxing problem that precludes direct application of classical communication techniques. Over the years several telemetry and communication systems have been proposed to overcome the problem of different underwater channel characteristics [63,64]. However, such methods utilise complex configurations and require powerful computational and instrumental capabilities that may not be compatible with portable power limited biotelemetry-processing requirements.

Underwater burst transmission alleviates the problem of the multipath transmission and allows a reliable underwater acoustic communication link especially in a shallow reverberant channel. The short pulse (burst) transmission is advantageous in cases where it is possible to correctly extract information from the signal simply utilising a transmission pulse interval that is always greater than the reverberation decay time.

It is well known that by assuming that the multipath reflects half way between the transmitter and receiver and that geometrical spreading from the point source is spherical, the time delay τ_d between the direct path and the multipath is given as:

$$\tau_d = \frac{\left[2*\sqrt{\frac{l^2}{4}+h^2}\right]-r}{c}$$

Where:

c is the speed of propagation of sound in water (typically 1500 m s^{-1}),
r is the transmitted distance,
h is the depth of the transducers. beneath the water surface.

For the current system, the tests were carried out in a large indoors underwater tank in Loughborough University, Leicestershire, UK with dimensions (9m long x 5m wide x 2m deep). Hence, if we assume that if the depth of the transducers below the water surface h was approximately 1m, and the average distance from the transmitter r was 4.5m, then τ_d= 282 µs.

From this analysis it can be seen that the first multipath occurs approximately 280 µs after the direct path in the specified underwater channel. This represents approximately a burst of three bits at a data rate of 10 kbit/s. From the experimental test, it was concluded that the multipath interference in the test tank was very severe, and nearly of the same magnitude as the direct path signals especially if the water surface is smooth as was the case here. The measured reverberation decay time in the current experimental system was in the order of 20 msec for a one-bit burst transmission length of 100 µs. Thus, a high burst transmission rate with such small depth-to-range ratio (≤ 0.2) will not make it practical in the receiver to detect and separate the successive main reflections from the multipath reverberations, because of the overlapping of the direct and multipath reflections. To overcome this problem, the present system utilised a much slower but narrower burst transmission format, with a digitally modulated carrier frequency, that always allows the transmitted pulse interval between the successive bursts to be greater than the reverberation decay time and to allow the reverberations to die away before the next data bit is transmitted with minimum loss in the received physiological parameters transmitted.

However, it is also important to mention that the choice of the minimum pulse width also depends on the Q-factor of the transducers used and on the transmission frequency, since the pulsed sonar transmits a short burst consisting of several cycles at the sonar operating frequency.

The transmitted pulse width τ is given as [61]:

$$\tau = (N/f_0) \qquad\qquad 9$$

where N is the number of the cycles at the operating frequency, and f_0 is the operating frequency of the transducer. It can be seen that for a specified pulse width to satisfy this equation the Q-factor of the transducer should be $\leq N$ in order to allow for the built up of the acoustic envelope in the transducer. Given from the measured and calculated data in appendices (A) and (B) of the transmitting hydrophone and power parameters of N=10, f_0= 68.9 kHz, and Q=4.1.

Thus, from Eq.(9), τ= 0.145ms. However, the bandwidth of the transmitted pulse must be smaller than the bandwidth of the transmitting transducer to attain maximum power peak transmission and is given as:

$$BW_T = f_0/Q = 69.8/4.1 = 17 \text{ kHz}$$

Hence, $\tau \geq 1/17 \times 10^3 \geq 0.06$ms.

Considering these limitations, and from the practical tests carried to characterise the multipath problem in the tank, it was shown that the best compromise solution for burst transmission in this design was the practical choice of a transmission pulse width of 100 µs with a one bit burst transmission at a maximum rate of approximately 50 bit/s. This was suitable for the current application, and satisfies the measured Q of 4.1 and the transmission frequency of around 70 kHz of the hydrophone used. Additionally, a biotelemetry system requires only a slow transmission rate because of the slow rate of change of the physiological signals. For example, the maximum possible heart rate of a healthy person could reach 210 beat/min., the maximum breathing rate is only 15 breath/min.[65], and the maximum calibrated depth pulse rate in the present work is 20 bit/s . Thus, these pulse interval and bit rate characteristics are compatible with the functional requirements of the pulse modulation formats adopted for both the telemetry of parameters such as heart rate and breathing rate and the Electrocardiogram (ECG) transmission in the present system.

3.4 Data Encoding, Modulation/Demodulation and Decoding

The previous section presented the main acoustic design concepts for digital acoustic biotelemetry over short ranges in a shallow underwater channel. This section outlines the other functional aspects of the system. From the generalised architecture for the system shown in Fig.(1), the source encoding operation matches the data source bit stream to the modulation and improves the system's reliability by means of coding. The modulation operation on the other hand matches the encoder output to the characteristic of the transmitter and the acoustic underwater channel. The decoding at the receiver inverts the linear operations of the coding to retrieve the transmitted data. In this work, the emphasis will be on data encoding and modulation techniques suitable for acoustic and radio telemetry in general and for underwater acoustic biotelemetry systems in particular. The most commonly and well known used methods, known as *pulse modulation,* are [2-7]:
  (i) Pulse Width Modulation (PWM),
  (ii) Pulse Position Modulation (PPM)
  (iii) Pulse Code Modulation (PCM).

For PPM signals, it consists of serial pulses of fixed duration where the sample amplitudes proportionally displace the pulses from a selected time reference. For PCM encoding, it converts each data sample from the periodic analogue-to-digital conversion of the data into a number of pulses (a word) and these pulses represent a digital value to each sample. We will examine the use of these encoding methods for underwater biotelemetry application later.

In underwater acoustic telemetry, modulation methods have principally been amplitude shift keying (ASK) and frequency shift keying (FSK), although there has been some recent work as cited in earlier chapters in the book on phase and quadrature phase modulation (QPSK) [73]. However, phase modulation of a carrier is impractical in biotelemetry applications for the following obvious reasons:

(i) Signal to noise ratios better when 20dB are rarely achieved in such environment, so there would be very limited dynamic range.

(ii) The fluctuation of non-linear multipath propagation of the acoustic signal produces a combination of signal with randomly varying phase distortions and fluctuations at the hydrophone, largely because of the phase-tracking problem [74]. To overcome this a very complex hardware set-up utilising adaptive tracking and estimation methods. The FSK approach is a relatively simple, low-performance form of digital modulation and is useful for telemetry transmission of low-rate digital data for up to 1500 bits/s in open water [75,76]. However, such a method is still prone to reverberation problems, namely, the constructive and destructive interference of the overlapping signals affect the identification of the binary bits. Also there is a requirement for two carrier frequencies, hence the energy required to transmit a data word is high, since every bit requires a separate pulse transmission. This is not compatible with the efficient and limited energy requirements of a biotelemetry system. Demodulation of the FSK signals is also complicated by problems of maintaining time synchronisation, as this requires self-synchronisation to extract the transmitted codes.

The operation of a binary shift keying (BFSK) modulator in a shallow water environment does not match with the slower bit rate transmission and the minimum bandwidth considerations of BFSK in the receiver. Because binary FSK is a form of frequency modulation, the formula for *modulation index (MI)* used in FM is also valid for BFSK and is given as [70]:

Where;
f_m is the mark frequency of the transmission.

$$MI = \frac{|fm - fs|}{fb}$$

f_s is the space frequency of transmission.
f_b is the input bit rate

Now let us consider the present severe multipath channel and the requirements of the current design example. The mark/space frequencies are 70kHz and 68kHz, dictated by the bandwidth characteristics of the projector and hydrophone, to prevent mutual interleaving in the receiver.

Additionally, the minimum bandwidth for BFSK is dependent on the modulation index. Consequently, in BFSK the modulation index is generally kept below 1.0 (between 0.5 and 1.0), so either two or three sets of significant side frequencies are generated.

Thus, for $MI = 0.5$, Eq.(10) yields theoretically:
$f_b = 2000/0.4 = 4000$ bits/s

This bit rate, although considered adequate in open water digital data acoustic telemetry applications. This is usually not practical for the bit rate requirements and receiver considerations for a multichannel biotelemetry system especially in a severe multipath environment. The difficulties with underwater data transmission using ASK and particularly OOK in reverberant environments have been previously recognised and studied [77,78]. However, ASK systems are considered simpler to implement and require less hardware for the bit transmission and detection synchronization compared to the other approaches and it has its advantages in this specific application. They also provide several unique interrogation codes depending on the frequency bands and the number of channels. ASK is also more power efficient, as it requires a single carrier transmission, making it more suitable for biotelemetry applications. However, the approach is still susceptible to multipath reverberations. Nevertheless, the system performance and reliability can be considerably improved by proper detection design in the receiver and greater transmitter power.

3.5 Physiological parameters adaptable to underwater biotelemetry

Physiological signals suitable for underwater biotelemetry are divided into two types [80-83]:
(i) bioelectrical variables e.g. electrocardiogram (ECG), electroencephalogram (EEG);
(ii) Other physiological variables such as blood pressure, respiration rate, and skin temperature of divers.

In addition, certain physiological parameters are of interest, notably depth and water temperature, because these are interrelated with the bioelectrical and physiological variables. In (i) the signals are obtained directly in electrical forms, whereas in (ii) (with the exception of respiration rate) they are measured as variations of resistance, inductance or capacitance.

Since the early work on underwater biotelemetry, the ECG and heart rate have been considered the most appropriate parameters for underwater

106   Chapter 3

transmission. This is mainly due to the clinical importance and the relative simplicity of the transducers and sense electrodes involved. In this chapter we emphasise on the combination of heart rate, respiration rate and depth parameters, selected because they enable a suitable correlation to be made between the human physiological condition and underwater activity. However, for the detailed classification of sensors and the physiology of bioelectrical events can be found in several biomedical instrumentation references [80-83].

In summary, for the design of a general purpose underwater acoustic biotelemetry there is no clear-cut way to employ either of the data encoding and/or the modulation and multiplexing techniques discussed in this chapter. This is because, as with many other systems, compromises must be made between the functionality of the design and the reliability, size and power consumption requirements. The flexibility desired for a low-cost research instrument can be achieved by reserving the solution to this compromise to the user, either through software control or simple hardware changes.

4. GENERAL DESCRIPTION AND OVERALL BIOTELEMETRY SYSTEM

This section presents the modular design methodology of the multichannel underwater biotelemetry system based on a modular design approach for a set of hardware modules of the selected physiological parameters and the associated software. Fig.(2) shows a block diagram of the underwater biotelemetry system architecture. It consists of a central micro controller-based chip in the transmitter and receiver sets with a modular electronic blocks (subsystems) for the different sensor interfaces interconnected to the central processor to define the telemetry protocol under the synchronised software control. The system is designed with a flexible modular approach and reprogrammable structure to accommodate for the telemetry link of a range of important physiological and diving parameters such as ECG, breathing rate , depth, body and water temperature. It can also be easily reconfigured for use for other medical and diving parameters.

This structure meets both the acoustic biotelemetry design criteria discussed in the last section and provides a low-cost general-purpose platform for fast prototyping of underwater biotelemetry research. In this work an 8-bit single chip microcontroller processor serves as the central function-processing unit for the multi-sensor interface and the individual signal conditioning circuits. It multiplexes the outputs and/or encodes the physiological data as appropriate to the user requirements, and to a format compatible with different underwater environment communication considerations. The appropriate communication capabilities and the biotelemetry protocols were determined by examining the several biopotential data and physiological signals selected for the current application. The biotelemetry of the multichannel biopotential pulse rate and depth data multiplexing from each sensor is accomplished via an 'inverse' data rate-interrupt multiplexing hierarchy in which the processor forms an 'intelligent' interrogator with each biopotential module linked to it. The interrupt hierarchy architecture is linked in real-time under a software interrupt priority control allocated for each signal. An identically synchronised software and interrupt priority control structure is implemented in the receiver system for maximum timing accuracy and minimum transmission error. Fig. (3) shows the relevant software structure of the multichannel programmable biotelemetry system.

The hardware elements of the analogous ECG biotelemetry is achieved via a microcontroller-based QRS identification, PCM digital data encoding algorithms and transmitted using (OOK) modulation methods. The same sequence of inverse pulse decoding with the appropriate digital-to-analog conversion (DAC) and filtering of the received signal is implemented in the receiver sub-system to retrieve the transmitted signal. This flexible and programmable hardware and software interaction under synchronised central microprocessing units in the transmitter and the receiver sub-systems outlines the new intelligent prototype underwater digital biotelemetry architecture presented here. The details of the transmitter/receiver hardware and relevant software design philosophy and implementation issues are presented elsewhere [84,85,86,87,88]. The main elements of the system are described here for completeness.

The two major telemetry tasks assigned to the system are :
1- Multichannel telemetry of the digital data.
2- The encoding/modulation of the continuos physiological signal in a format suitable for digital telemetry such as ECG transmission.

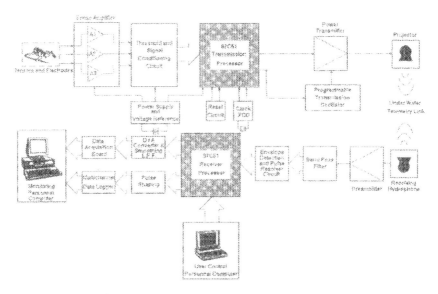

Fig.2 Functional block diagram of the programmable underwater biotelemetry system

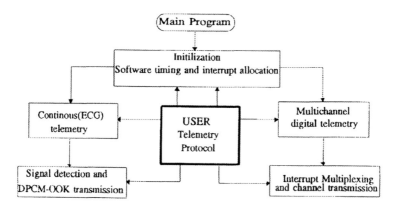

Fig.3 Software structure of the programmable underwater biotelemetry system

4.1 - Programmable hardware and software design methodologies

Figures (2) and (3) show the integrated software and modular hardware design concepts of the system. The modular hardware and programmable design of the system provide a flexible system suitable for telemetering different physiological signal and in different formats and also allow interchanging between any of the above transmission formats (intermittent or continuous physiological channels) according to the monitoring requirements. It should be noted that minimal software and hardware changes are required for additional sensors or different coding schemes. The development of such an embedded control system requires a specific software design.

The overall control, computational tasks and algorithmic calculations that instigate the two transmission formats are accomplished by the relevant 'assembly software programming' written for the central 87C51 processor (in this case) and dedicated for each task. The assembly programming approach is ideal for such bit level work, offering the advantages of high execution speed and determination of the exact timing involved in the software development, in addition to the availability of the emulation facilities at the test site in Loughborough University. This centralized single-chip functional control offers a highly flexible and versatile configuration due to the compatibility of the hardware with the other transmitter modules, and allows a choice of transmission mode without any hardware alterations in the signal acquisition and transmission modules.

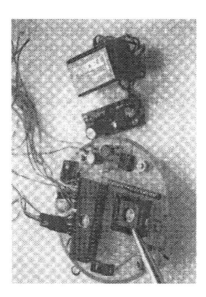

**Fig. 4-a Biotelemetry transmitter showing the
Intel 87C51 microcontroller and transmitter power unit.**

Both the transmitter and receiver contain identical 8-bit single-chip Intel 87C51 mcirocontrollers; see Fig.(4-a).The complete transmitter is encapsulated in a pressure-proof housing carried by the diver inside the dry suit; see Fig.(4-b).

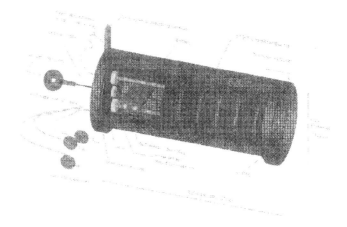

Fig.4-b The pressure-proof housing of the biotelemetry system.

Different physiological signals are acquired form the appropriate sensors; in case of the ECG, waterproof skin electrodes are attached to the diver's chest and the associated connectors are passed through the housing.

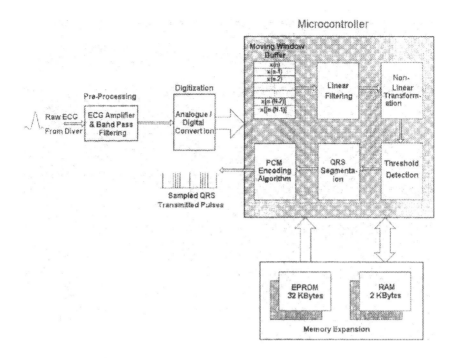

Fig.5   Real-time QRS detection and telemetry transmission.

Fig. (5) illustrates the algorithmic concepts used for the ECG detection and telemetry mechanisms mentioned earlier.

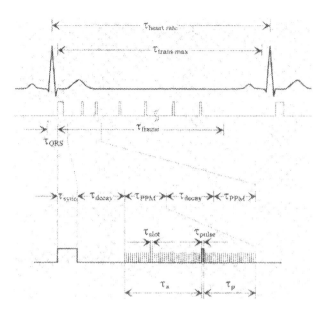

Fig. 6   ECG digital PPM encoding ad transmission format

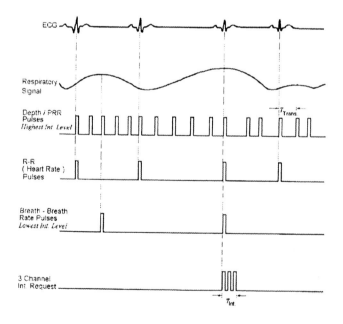

Fig.7   Multichannel Interrupt-based biotelemetry timing diagram

Fig. (6) illustrates the corresponding ECG digital PPM data encoding and transmission formats and Fig (7) illustrates the timing diagram of the multichannel interrupt-based physiological and diving data transmission structures.

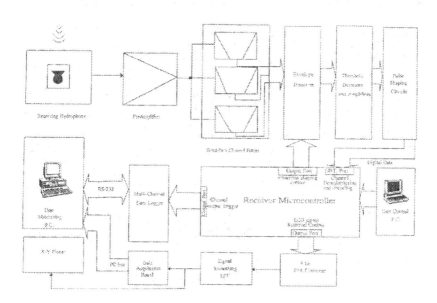

Fig. 8   Block diagram of the receiver system

Fig. (8) shows the block diagram of the microconroller-based receiver system that is designed with the matching software protocols compatible with the transmitter system.

4.2 Experimental results and discussion

In this section we present some of the experimental results of the performance of the programmable system discussed earlier. The main objective of the tests was to verify the functionality of the system. The system was tested successfully after careful underwater and depth proof encapsulation on different SCUBA divers in a large indoor underwater tank (9 x 5 x 2 m) in Loughborough University, UK. This test arrangement is shown in Fig. (9).

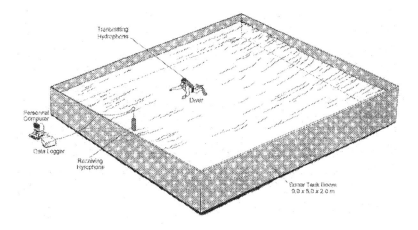

Fig. 9   Underwater Biotelemetry diver testing environment.

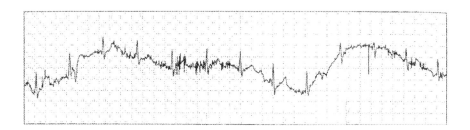

Fig.(10)   Telemetered ECG signals for the tank test.

Fig. (10) shows a typical telemetered ECG data received by the system.

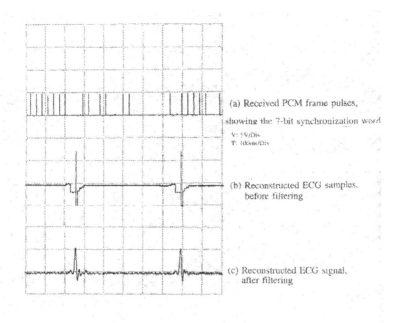

Fig. 11   Received PCM pulses and reconstructed ECG signals

Fig. (11) illustrates the results of the transmitted PCM encoded ECG samples including the 7-bit synchronization word used to distinguish between successive ECG samples. It also shows the corresponding received ECG samples before filtering and the final reconstructed ECG signals. Three channels of data were successfully observed and distinct ECG signals were received during the exercise phases of the trials. For the multichannel recordings, interrupt-based heart rate, breathing rate and depth measurements were telemetered form the system.

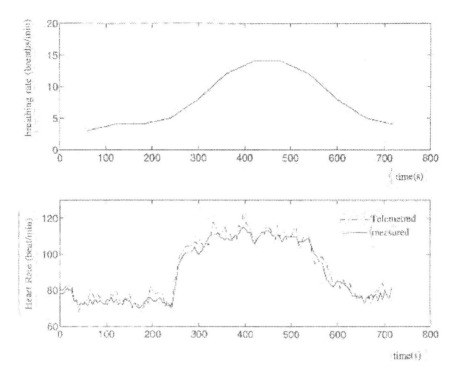

Fig. 12 Comparative results of telemetered heart rate and breathing rates with measured heart rate signals.

Fig. (12) shows the results of the acoustically telemetered heart rate and breathing rates captured by special data logger in the receiver system (Grant/Elteck, Model 1200, Cambridge, and UK) via an RS-232 to a portable laptop PC. The diver were also equipped (under their dry suits) with special commercial heart rate monitoring devices (Polar Electro Sport Tester, Model-4000, Kemple, Finland) for the purpose of comparing performances between the measured heart rates and telemetered heart rates during diving and underwater swimming. The instrumented SCUBA diver was asked to swim in the tank performing various manoeuvres. For the ECG telemetry trial, the quality of the recording was sustained during the diving exercises with clear QRS (main clinical heart signal features) in the electrocardiogram signal and observable during the tests as shown.

The important measures established by the system's tests are the comparative evaluations between the PCM and PPM encoding schemes sued of the ECG transmission and the performance measures between the telemetered and measured heart rate signal during different diving and underwater swimming manoeuvres. The results of these tests are summarised in tables (1) and (2) respectively.

		PCM			PPM	
N (bits/s)	τ_{decay} (ms)	$f_{heartrate}$ (beat/min)	f_{bit} (s^{-1})	$f_{heartrate}$ (beat/min)	f_{bit} (s^{-1})	
8	20	35	50	108	175	
7	20	40	50	140	213	
6	20	46	50	165	227	
5	20	54	66	180	215	
8	15	46	66	118	197	
7	15	52	66	158	251	
6	15	60	66	191	279	
5	15	70	66	212	274	

Table (1) Comparative performance of PCM and PPM for different quantization levels (N) and reverberation decay times (τ_{decay})

The results from table (1) demonstrate that the performance of the system for continuos ECG transmission depends on the digital encoding method used and the reverberation decay time inherent in the relevant underwater channel. It can also seen that for the same quantization level and reverberation delay, PPM encoding provides ECG signals at higher bit rates and consequently can operate at higher bit rates compared to the PCM method. This is particularly useful when transmitting the ECG at higher heart rates, such as during hard underwater work or fast swimming required in environments such as in off shore installations.

The results of table (2) show that the absolute percentage error correlation between the telemetered heart rate data and the data during the dive was ± 3.3%. However, such percentage is acceptable as most commercially available heart rate monitors have a loss in accuracy between 1% and 3% at higher heart rates (> 150 beat/min). The depth variation is not shown, as the average depth was constant at approximately 4ft(1.2m) logged by the pressure/depth channel at 635 ppm and due to the shallowness in the tank.

The maximum percentage error referred to the calibrated data was <1% at the maximum test depth.

Diving Regimes	Average Heart Rates (beats/min ± S.D.)	
	Measured	Telemetered
Lying motionless	74± 2	76± 3
Circular swimming	106± 8	109± 7
Follow-up lying still	89± 10	91± 11

Table (2) Comparative performances of the telemetered and measured heart rate data

This error could be attributed to the logging process, or slight variations in the sensor's component circuitry and ambient temperature variations and owing to the fact that the sensor is an integral part of the signal converter circuitry since it converts resistance variations into frequency variations. Relevant heart rate variability studies that has been carried out for further physiological analysis during diving from data acquired using the current system are presented elsewhere [89]. Also some of preliminary studies integrating this system with the use of Artificial Neural Networks (ANN) for the detection of the likelihood of medical disorders during diving are given in [90]. Other ANN applications in underwater signal processing are also detailed in the following chapters of the book.

5. CONCLUSIONS

This chapter presents a microcontroller-based integrated programmable underwater acoustic biotelemetry system. The main aim of the work is to present the design concepts of such system suitable for real-time monitoring of a free swimming SCUBA diver or swimmer in a reverberant underwater environment.

The design procedures presented in this work provide a modular hardware and software organization for the transmitter and the receiver. The microcontroller-based architecture provides the necessary time-base synchronization and trade-off between the acoustic telemetry limitations and the corresponding physiological monitoring. This versatile modular approach allows the digital telemetry of data such as heart rate, breathing rate and depth and/or analogue signals such as the ECG. It also allows the desired mode of transmission to be selected and the addition of further monitoring channels without any alteration to the existing hardware and only minor modifications to the system's software.

The recent advances in DSP architectures and application-specific IC's (ASIC) technology could make the possibility to improve the performance and miniaturise the system further in several ways. For example using surface mounted components with a redesigned housing could allow further miniaturisation of the biotelemetry system. Using alternative single-chip processors such as the 16-bit Intel-80C196 series could improve the performance of the system with modified specifications such as increased speed (0.66 µs), higher vectored interrupt levels (16 levels), and on-chip 10-bit A/D converter with sample/hold. Another solution is to use a newly developed 8-bit microcontroller based on the 'Smart Power' technology where the single processor IC (H081-SGC-Thompson) incorporates a 60-V DMOS FET H-bridge rated at a continuous current of 1A. This technology makes possible the development of programmable output power transmission features in a single chip. In the sensor and signal acquisition modules, future options include the use of the newly developed smart sensors using micromechanical system (MEMS) technology. These enables the measurement of different physical and physiological parameters with an in built processor fabricated with single-chip specific purpose IC. These technologies could lead to further integration and implementation of special purpose underwater biotelemetry system where the medical signals acquisition, proper coding and modulation techniques are integrated to be compatible with the operation of Third Generation (3G) telephonic systems for global communications coverage and compatibility.

Acknowledgements

The author is grateful to the Committee of British Universities Vice Chancellors and Principals for their ORS award and for the Armenian General Benevolent Union Grants Committee, London and Loughborough University for their partial support for this work.

6. REFERENCES

1. Mackay, R.S., (1970). *Biomedical Telemetry: Sensing and Transmitting Biological Information from Animals and Man*, 2nd Ed., New York: John Wiley.
2. Kimmich, H.P. (1975). 'Multichannel Biotelemetry-Review', Biotelemetry, 2, 207-255.
3. Jeutter, D.C., (1983). 'Overview of Biomedical Telemetry Techniques', IEEE Eng.In Med.& Biol., 2, 17-24.
4. Kimmich, H.P., (1980) 'Artifact free measurement of biological parameters, biotelemetry a historical review and layout of modern development, In Handbook Biotelemetry and radio tracking, AMLANER, C.J. and MACDONALD, D.W. (eds.), Pergamon press, N.Y., 3-20.
5. Fryer, T.B. and Sandler, H. (1974), 'A Review of Implant Telemetry Systems', Biotelemetry, 1(6), 351-374.
6. Ysenbrandt , H.J.B.,Selten, A.J., Koch,T., Kimmich, H.P., (1976). 'Biotelemetry-Literature Survey of The Past Decade', Biotelemetry,3, 145-250.
7. Meindel ,J.D., and Ford, A.J. (1984). 'Implantable telemetry in biomedical research, IEEE Trans. Biomed. Eng., 31, 817-821.
8. Adrian, E.D. and Ludwig, C. (1938). 'Nervous Discharge from the olfactory organs of Fish', J. Physiol., 94, 441-460.
9. Slater, L.E. (ed.) (1965). *Biotelemetry, the use of telemetry in animal behaviour and physiology in relation to ecological problems*, Pergamon press, New York.
10. Priede, I.G (1980). An analysis of objectives in telemetry studies of fish in the natural environment', In Handbook on biotelemetry and radiotracking, AMLANER, C.J. and MACDONALD, D.W.(eds.), Pergamon press, NewYork, 105-118.
11. Frank, H.T. (1968). 'Telemetring of Electrocardiogram of Free Swimming Fish (*Salmo Irideus*)', IEEE Trans. Biomed. Eng., 15, 2, 111-114.
12. Scidmore, A.K., Beyer, J.B., Weiss D., and Guckel, H. (1976). 'EEG Monitoring of Free Swimming Fish, Biotelemetry, 3, 227- 230.
13. Kudo, Y., Urda, K. (1980). 'The measurement of the electrical activity of the olfactory bulb in free swimming carp by underwater telemetry system', IEEE Trans. Biomed. Eng., 27, 694-698.
14. Wheatherley, A. H., Rogers., S.G., Pincock, D.G., and Patch, J.R.(1982). 'Oxygen consumption of active rainbow trout derived from electromyograms obtained by radiotelemetry', J. Fish Biol., 20, 479-498.
15. Trefethen, P.S.,' (1956).'Sonic equipment for tracking individual fish', U.S. fish & Wildlife Serv. Spec. Sci., Rep. No.179.

16. Stasko, A.B. (1975). ' Underwater biotelemetry,an annotated Bibolography', Canadian Fish. Mar. Serv. Resour. Dev. Branch Marit. Tech. Rep. Ser. MART-534, Canada, 31pp.
17. Henderson, F. G. and Hasler, A. D. (1966)' An Ultrasonic Transmitter for use in the Studies of Movement of Fishes', Trans. Am. Fish.Soc., 95, 350-356.
18. Young ,A.H.,Taylor,P., and Holliday, F.G.(1972).' A small sonic tag for the measurement of locomotor behaviour in fish,' J.Fish Biol., 4., 57-65.
19. Monan, G.E., and Thorne, D.L. (1973). 'Sonic tags Attached to ALaska King Crab', Marine Fisheries Review, 37(2), 9-15.
20. Carey, F.G., and Lawson, K.D.(1973).' Temperature regulation in free swimming bluefin tuna, Comp. Biochem. Physiol., 44, 375-392.
21. Rubinoff, I., Graham, J.B., and Motta J.(1986). 'Diving of the Sea Snake (*Pelamis Planturus*) in the Gulf of Panama: I. Diving Depth and Duration', Mar. Biol., 91, 181-191.
22. Wilson, R.P., Grant, W.S., and Duffy, D.C. (1986).'Recording devices on free-ranging marine animals', Ecology, 67, 1091-1093.
23. Priede, I.G., and Young, A.H. (1977). 'The ultrasonic telemetry of cardiac rhythms of wild brown trout as an indicator of bio-energies and behaviour, J. Fish Biol., 10, 229-318.
24. Bottoms, A. and Marlow, J.(1979).' A new Ultrasonic Tag for the Telemetry of Physiological Functions from Aquatic Animals', Mar. Biol., 50, 127-130.
25. Wardle, C.S., and Kanwisher, J.W. (1973). 'The significance of heart rate in free swimming Cod (*Gadus Morhua*): Some observations with ultrasonic tags, Mar. Behav. Physiol., 2, 311-324.
26. Rogers, S.C., Weatherley, A.H., Pinckock, D.G., and Patch. J.R. (1981). 'Telemetry electromyograms and oxygen demands of fish activity, Proc. 3rd. Int. Conf. Wildl. Biotelemetry, Univ. of Wyoming, Laramie, 141-150.
27. Stasko, A.B., and Horrall,R.M. (1976). 'Method of counting the tail-beats of free swimming fish by ultrasonic telemetry', J. Fish Res. Board. Canada, 33, 2596-2598.
28. Ferrel, D.W., Nelson, D., Standora, T.C., and Carter, H.C. (1974).'A Multichannel Ultrasonic Biotelemetry System for Monitoring Marine animal Behaviour', Trans. ISA , 13,120-131.
29. Bjordal, A., Floen, S., Todland, B., and Huse, I.(1986) ' Monitoring biological and environmental parameters in aquaculture', Model., Ident. and Control, 7,209-218.
30. Kanwisher, J., Lawson, K., and Saunders, G. (1974).' Acoustic telemetry from fish', Fish. Bullet., 72, 251-255.
31.Braithwaite, H. (1974) 'Some measurements of acoustic conditions in rivers', J. Sound & Vib., 37, 557-563.

32. Meister, R., and ST. Laurent, R. (1960) 'Ultrasonic absorbtion and velocity in water containing algae in suspension, 'J. Acous. Soc. Amer., 32, 556-559.
33. Brumbaugh, D. (1980),'Effect of thermal stratification on range of ultrasonic tags', Underwater telemetry newsletter, 10, 1-4.
34. Urick, R.J., (1975), *Principles of underwater sound for engineers'*, 2nd. Ed., McGraw-Hill:N.Y.
35. Burt, E.G., and Rigby, L.(1985),'Electromagnetic through-water communication' J. Soc. Underwater Tech., 11, 14-18.
36. Slater, L.E. (ed.), (1965) *'Biotelemetry'*, Oxford: Pergamon Press.
37. Magel, J.R., Mcradle, W.D., and Glaster, R.M.(1969) 'Telemetred heart rate response to selected competitive swimming events', J.Applied Physiol., 26, 764-770.
38. Baldwin, H.A.(1965),'Marine biotelemetry', Bioscinece,15, 95-97.
39. Oka, Y., Utsyama, N., Koda, K., and Kimura, M. (1963) 'Studies of telemetring on ECG and respiratory movements during running, jumping and swimming', Med. Elec. Biol.Eng., 1,574-579.
40. Tucker, D.G., and Bazey, B.K. (1966), ' *Applied underwater acoustic'*, pergamon press:London.
41.Anderson,V.C. (1970).'Acoustic communication is better than none', IEEE Spectrum, 63-68.
42. Frampton, C., Riddle, H.C., and Roberts, J.R. (1976) 'An ECG telemetry system for physiological studies on swimmers', Biom. Eng., 11, 87-90.
43. Frampton, C., Riddle, H.C., and Roberts, J.R. and Watson, B.W. (1974), 'An inductive loop telemetry for recording ECG of a swimmer', J. Physiol., 241,14-16.
44. Glaser,R.M., and Mcradle, W.D. (1969) 'A Radiotelemetry Transmitter for Monitoring Heart rate of Human Engaged in Physical Activity', The Res. Quarter. of The American Assoc. of Health,Physic. Educ. and Recreation, 40(3), 640-642.
45. Jacob,R., Ridde, H.C., and Watson, B.W. (1973)' Circuits for searching a signal from a three aerial system during inductive loop telemetry', Biomed. Eng.,8, 295- 297.
46. Shcultz, C.W. (1970) 'Underwater communication using return current density', Proc. IEEE, 1025-1026.
47. Zweizig, J.R., Adey, W.R., Hanley, J., and Cockett, A.T. (1972) 'EEG monitoring of a free swimming diver at a working depth of 15m', Aerospace Med., 43, 403-407.
48. Utsuyama, U., Yamagushi, H., Obara, H., and Miyamato, H.(1988) ' Telemetry of human electrocardiogram in aerial and aquatic environments, IEEE Trans. Biomed. Eng., 35., 882-884.
49. Weltman, G., and Egstrom, G.H. (1969) 'Heart rate and respiratory response correlation in surface and underwater work', Aerospace Med., 40, 479-483.

50. Thalman, E.D., Sponholtz, D.K., and Lundgren, C.E. (1976)'Chamber-based system for physiological monitoring of Submerged exercising subjects, Undersea Biomedical Res., 5, 293-300.
51. Lovik, A., Lloyd, J., and Tangen, J.M. (1983) 'A general system for the physiological diver monitoring', Proc. IEEE Ocean'83 Conf., 1, 464-466.
52. Pilmanis, A.A., Given, R.R., and Pilmanis, V.M. (1971) 'Physiological and biological studies from The Hydrolab', Hydrolab J. , 1, 30-42, 1971.
53. Jung, W. and Stolle, W., (1981) 'Behaviour of Heart Rate and Incident of Arrhythmia in Swimming and Diving', Biotelemetry, 8, 228-239.
54. Slater, A., and Bellet, S. (1969)'An underwater temperature telemetring system', Med. Biol. Eng., 7, 633-639.
55. Slater, A., Bellet, S., Kilpatrick, D.G. (1969), 'Instrumentation for telemetring the electrocardiogram from Scuba Divers', IEEE Trans. Biom. Eng., 16,148-151.
56. Kanwisher, J., Lawson, K., and Strauss, R. (1974). 'Acoustic telemetry from human divers', Undersea Biom. Res., 1, 99-109.
57. Skutt, R.H., Fell,R.,and Hagstrom, E.C. (1971). A Multichannel ultrasonic underwater telemetry system', Biotelemetry, 1, 30-38.
58. Clay, C.S., and MedwinE, H. (1977). *'Acoustical Oceanography: principles and applications'*, Wiley;N.Y.
59. Berkhovskish, L. and Lysanov,Y. (1982). *'Fundamentals of ocean acoustics'*, Springer-Verlag:N.Y.
60. Flatte, S.M. (ed.) (1979)' *Sound transmission through a fluctuating ocean'*, Cambridge Univ. Press, Cambridge:U.K.
61. Burdic, W.S. (1991). *'Underwater signal analysis'*, Prentice-Hall, 1991.
62. Jensen, F.B., and Kuperman, W.A. (1983).'Optimum frequency of, propagation in shallow water environments', J. Acous. Soc. Amr., 73, 813-816.
63. Baggeroer, A.B. (1984). 'Acoustic telemetry-An overview', IEEE J. Ocean. Eng., 9, 229-235.
64. Coates, R., and Williams, P. (1987).'Underwater acoustic communications: A review and bibliography', Proc. Inst. Acoust., UK.
65. Ray, C.C. (ed.) (1974),' *Medical Engineering'*, Rev., Year book Medical Publishers, Chicago.
66. Jacobs, I. (1974).'Practical application of coding', IEEE Trans. Inf. Theory , 20, 305-310.
67. Pieper, J.F., Proakis, J.A., and Wolf, J.K. (1978).' Design of efficient coding and modulation for a Rayleigh fading channels', IEEE Trans. Inf. Theory, 24, 457-468.
68. Proakis, J. (1983).' *Digital communications'*, McGraw-Hill:N.Y.
69. Viterbi, A.J, and Omura, J.K.(1979).*'Principles of digital communication and coding'*, McGraw-Hill:N.Y., 1979.
70. Tomasi, W. (1992). *'Advanced electronic communication systems'*, 2nd. ed.,Prentice-Hall:N.J.

71. Fryer,T.B.(1974) 'A Multichannel EEG telemetry utilizing a PCM sub-carrier',Biotelemetry,1, 202-206.
72. Hull, M.L., and Motec,C.D.(1974).' PCM telemetry in ski injury research, Biotelemetry, 1, 182-184.
73. FalahatiA, A., Woodward, B. and Bateman, S.C. (1991).'Underwater channel models for 4800 b/s QPSK signals', IEEE J. Ocean. Eng., 18, 12-20.
74. Mohanty, N. (1987). ' Phase tarcking error in a fading channel', Proc. Int. Fed. Telemetry Conf.(IFT), 203-205.
75. Catipovic, J.A. (1990).' Performance limitation in underwater acoustic telemetry', IEEE J. Ocean. Eng., 15, 205- 216.
76. Brock, D.C., Bateman, S.C., and Woodward, B. (1986)' Underwater acoustic transmission of low-rate digital data', Ultrasonics, 24, 183-188.
77. Andrews, R.S., and Turner, L.F. (1977). ' On the performance of underwater data transmission system using amplitude-shift-keying techniques', IEEE Trans. on Sonic and Ultrasonic, 23, 64-71.
78. Andrews, R.S., and Turner, L.F. (1976) 'Investigation of amplitude fluctuations of high frequency short-duration pulses propagated under short-range shallow water environment', J. Acoust. Soc. Amer., 58, 331-335.
79. Carden, F. (1990)' Design parameters for FM/FM system', Proc. Int. Telemetering Con. ITC/90, 241-247.
80. Cromwell, L.,and Weibeli,F.J.(1980). 'Biomedical instrumentation and measurement', Prentice-Hall:N.J.
81. Tompkins, W.J., and Webester, I.G. (eds.) (1982).' *Design of microcomputer-based medical instrumentation*', Prentice-Hall:N.J.
82. Cobbold, R.C. (1974).'*Transducers for biomedical measurements; principles and applications*', Wiley:N.Y.
83. Geddes, L.A., and Baker, L.E. (1989). '*Principles of applied biomedical instrumentation*, 3rd ed., Wiley:N.Y.
84. Istepanian,R.S.H. (1994).' *Use of microcontrollers for diver mointoring by underwater acoustic biotelemetery in multipath environments*', Ph.D. dissertation, Loughbrough University, UK.
85. Woodward, B. and Istepanian, R.S.H. (1993) 'Acoustic Biotelemetry of data from divers', Proceedings 15th. IEEE Annual International Conference of Engineering in Medicine and Biology, San Diago,USA, 1000-1002.
86. Woodward, B. and ISTEPANIAN, R..H. (1995). ' The use of Underwater Acoustic Biotelemetry for monitoring of ECG of a swimming patient', Proceedings of 1st. Regional Conf. IEEE Engineering in Medicine and Biology Society , New Delhi, India,4, 100-108,1995.
87. Istepanian, R. and Woodward, B. (1997).'Microcontroller based ECG underwater telemetry system', IEEE Trans. Information Technology in Biomedicine, 1,2, 150-154.
88. Istepanian, R.S.H, and Woodward, B. (1996)' Underwater acoustic telemetry of complex analogue signals in a multipath channel' , In *Subsea Control and data Acquisition* (Andrianssen,L., Phillips, R., Rees C., and

Cattanach J. (eds.), Mechanical Engineering Publications Ltd., London, 171-18.

89. Istepanian, R.S.H. and Woodward, B. (1996). 'Spectral Analysis of Heart Rate Variability during Scuba Diving', Proceedings 18th. IEEE Annual International Conference of Engineering in Medicine and Biology, Amsterdam, Holland, 430-432.

90. Istepanian, R.S.H.and Woodward, B.(1997).' Use of Neural Networks in Telemedical Underwater Monitoring', Journal of Telemedicine and Telecare,3, Suppl.1,70-72.

4

Digital underwater voice communications

H Sari and B Woodward
Department of Electronic and Electrical Engineering, Loughborough University, LE11 3TU, UK

1. INTRODUCTION

In many underwater activities, a voice link is an essential requirement for divers [1-9]. Although advanced digital systems for underwater acoustic data transmission and reception have been devised, this is not the case for through-water voice communications. Most commercially available systems use outdated analogue technology and therefore suffer practical limitations compared with the sophisticated digital telecommunications systems now available above the surface. One advantage of a digital system is that it enables private communication links between divers or between divers and the surface so that there is no unwanted cross-talk. Another advantage is that it can be used to transmit scientific data as well as voice information. In this chapter, we consider a new design methodology that includes the implementation of speech signal processing algorithms and the transmission and reception of speech in digital format. The main consideration in the design is to provide a diver with a comparable level of communications capability as is provided by a digital mobile telephone.

1.1 Multipath Underwater Propagation

The most difficult problem associated with underwater acoustic communication channels is the multipath propagation of limited bandwidth signals, which depends on the location of the transmitter and receiver, reverberation from boundaries and volume reverberation [10-16]. Shallow channels in which divers operate are generally considered to be extremely hostile for high data rate communication, because they exhibit signal dispersion and phase fluctuation. Multipath propagation can therefore cause severe degradation of the transmitted signals. The combination of high data rate and multipath propagation can generate inter-symbol interference (ISI) and an important figure of merit is the multipath spread in terms of symbol intervals. Typical multipath spreads in a horizontal underwater acoustic channel may be tens or even hundreds of symbol intervals for moderate to high data rates [17,18]. Achieving a high data rate is the most difficult task in designing a digital communication system [18-22].

The design of modern data transmission systems has been accompanied by extensive research to achieve high data rates by channel equalisation [18,19], array processing [23], phase steering and adaptive beam forming [20,21]. Although these methods may be beneficial in communications applications for autonomous underwater vehicles, neither array processing nor adaptive beam forming is suitable for diver communications due to the increased signal processing complexity and physical size of the system. In the prototype system presented here, the transmitter and receiver are mobile; therefore omnidirectional transducers are used. No ISI suppression method is adopted but techniques are suggested for future designs.

1.2 Speech in Underwater Environments

The main factors influencing the efficient transmission of speech underwater are noise and distortion, which are introduced at the source, in the channel and at the receiver. Distortion introduced at the source depends on the type of microphone employed, the shape of the diver's mask and the breathing gas constituents. In addition, bubble noise and breathing noise, which must ideally be cancelled or at least minimised before transmission, have a significant effect on speech quality. Several noise-cancelling methods have been applied to enhance speech signal quality whilst leaving the diver's speech undistorted.

Two major components in any underwater voice communication system are microphones and earphones [5]. The microphone must resist water damage and corrosion and it must not enclose any gas that can be compressed and decompressed because when placed inside the diver's mask it is subjected to the full ambient pressure. In the case of diver breathing helium, the gas increases the frequency range of human speech up to around 8-10 kHz [26,27], which is well above the response range of ordinary microphones. Adequate steps must be taken either to pressure proof or pressure balance the microphones, as the frequency response at atmospheric pressure is likely to be adversely affected at elevated pressure [28].

1.3 Effect of Ambient Pressure on Speech Production

The combined effects of pressure, breathing gas constituents and the difficulty of normal vocalisation make for considerable problems in transmitting clear and intelligible speech. As depth increases, the pressure of inhaled gas must be equal to the ambient water pressure. At depths below about 30 m, i.e. at absolute pressures in excess of 4 bar, the overall sound pressure level of voiced sounds increases and a relative loss of spectrum intensity at high frequency occurs, resulting in speech having a nasal quality [27].

For depths greater than about 50 metres, divers usually breathe 'heliox' (a mixture of helium and oxygen) or 'trimix' (a mixture of oxygen, helium and nitrogen), the mixture varying as a function of depth. This inevitably results in a variation in the speed of sound, which directly influences the speech signal spectrum. Due to the increase in the speed of sound in helium, being nearly three times higher than in air, voice distortion occurs and the speech becomes garbled, often referred to as the "Donald Duck" effect [29]. To compensate for the effect of the helium environment and ensure intelligible speech [26, 29-33], several algorithms have been developed to correct only the formant frequencies of the helium speech, implementing both time domain processing and frequency domain processing. Helium speech correction has not been applied here but with a digital signal processor (DSP), it is possible to implement any of the helium speech enhancement algorithms.

1.4 Underwater Hearing

Underwater hearing thresholds define the minimum output power requirement of any diver-to-diver speech communication system. While much is known about the limitations of the ear in air, little is known about its limitations underwater. The audibility threshold performance of the human ear in water has been studied [34,35] and it appears that underwater hearing thresholds are 18-56 dB higher than those in air, the difference increasing with frequency. One of the general conclusions drawn from the studies is that the middle ear is not used in underwater hearing but is by-passed by the mechanism of bone conduction. Hence currently available underwater voice communication systems usually use piezo-ceramic bone conduction devices.

1.5 Speech Quality Enhancement from a Diver's Mask

In many speech communication systems, background interference degrades the quality or intelligibility of speech. In a diving environment the speech is often accompanied by extraneous noise, including breathing noise produced during inhalation by airflow through the regulator of a conventional SCUBA or by free-flowing air through a helmet or full-face mask. Due to its high signal amplitude, the breathing noise has a deleterious effect on the clarity of communications. There is also bubble noise, which is generated during exhalation, including during speech periods, by escaping air from the regulator. Although bubble noise occurs outside the mask cavity, a microphone inside the mask can detect it.

2. SYSTEM DESIGN

2.1 Introduction

Rapid development in mobile telecommunication technology has been made possible by advances in low bit rate speech coding methods and DSPs [36,37]. To achieve real-time, digital underwater acoustic voice communications, the speech signal must be processed, then transmitted and received by implementing a suitable digital modulation and demodulation technique. Microprocessors are inappropriate for such applications so DSPs are essential to implement the complex signal processing computations. A DSP allows the design of a physically small unit and allows for modifications in speech coding, modulation and demodulation algorithms. The system described here operates in two different modes. In transmission mode, it must be able to: (i) digitise an analogue speech signal; (ii) compress speech samples; (iii) encode compressed speech parameters; and (iv) transmit digital signals representing the encoded speech parameters. In receive mode, it must be able to: (i) receive digital signals and decode speech parameters; (ii) synthesise speech signals; and (iii) generate analogue speech signals.

2.2 System Architecture

The system structure is shown in Fig. 1. The transmitter consists of a keypad, a speech pre-processing circuit (preamplifier, anti-aliasing filters and analogue-to-digital converter), a power amplifier and a transmitting transducer. The receiver includes the same transducer, a preamplifier, band-pass filter, envelope detector and analogue-to-digital converter. At the speech output end, there is a post-processing circuit, which includes a digital-to-analogue converter, low-pass filter and amplifier. The DSP is a 32-bit TMS320C31 (33 MHz), from Texas Instruments, which is a so-called "third generation" device that is capable of executing 33.3 million floating point operations per second (MFLOPS). It is widely used for complex mathematical and fast signal processing applications. Speech signal conditioning and digital-to-analogue or analogue-to-digital conversion is achieved with an application-specific speech codec, a TLC320AC01 Analogue Interface Circuit (AIC), also from Texas Instrument, which has a synchronous, serial digital interface.

The DSP maintains complete control of the system and performs communication tasks with peripheral units. Its features include +5V single supply operation, two general-purpose 32-bit programmable timers and a serial port supporting a maximum transmission data rate of 4.1625 Mbit/s. Allocated registers for the timers and the serial port can be programmed to perform a wide variety of tasks, and the timers are effective in the generation of accurately timed, high-speed pulse sequences for the different pulse characteristics. Both the timers and the serial port use internal interrupts. The DSP also incorporates two general-purpose input/output ports, which can be operated under software control, and four external interrupt inputs that allow efficient communications with the external peripheral units.

The DSP has a 2k-word internal Static Random Access Memory (SRAM) and a 64-word program cache memory for maximum efficiency in program execution time. The on-chip SRAM can be used either to execute programs from slow speed off-chip memory devices or as temporary memory to store speech samples and relevant data. The 24-bit address bus enables expansion of the off-chip memory to a maximum of 16.77 Mword. In the current design, due to the limited size of the internal SRAM and the exclusion of an internal Erasable Programmable Read Only Memory (EPROM), off-chip memory devices are incorporated. Computationally complex speech processing, in terms of analysis and synthesis, and data transmission and receiving algorithms are downloaded into the internal SRAM of the DSP and executed there in order to improve the operational efficiency of the DSP.

Although four external peripherals are incorporated in the system, their number can always be increased if a modification in the design is required.

Communications with these peripherals is essential for the DSP to perform its operations. Some of the peripheral units, such as the EPROM and SRAM are continuously accessed. Some only instruct the DSP through its interrupt inputs to initiate communication. The last one enhances the system's operational speed since the DSP does not need continuously to access these units. The only drawback is that a limited number of peripheral units are allowed to exploit this feature of the processor, which is determined by the number of interrupts.

2.3 Transmit-Receive Operation

The system requires a suitable transducer that can convert electrical energy into mechanical energy or produce an electrical signal when excited by a pressure wave. A single omnidirectional electrostrictive transducer made from lead zirconate titanate (PZT) (type D-170, Universal Sonar Ltd., U.K), operating at its resonance frequency of 70 kHz and radiating spherically, is used as both projector and hydrophone. The electrical characteristics of the transducer are listed in Table 1. Two different electret microphones (MR-8406 and CF-2949, Knowles Ltd) were tried. During diving trials, it was found that the subjective speech quality achievable with the waterproof MR-8406 microphone dramatically decreases when the ambient pressure is increased and noise is generated. The CF-2949 microphone proved to be superior in terms of the clarity and intelligibility of speech because it is specifically designed to compensate for changes in ambient pressure. The choice of earphones is also important since these are also used underwater. Piezo-ceramic speakers can couple well into water and have a response of 500 Hz to 20 kHz.

The system is designed to operate in half-duplex mode. The diver controls its operation mode, either as a transmitter or as a receiver and this may be achieved by several different methods. Voice-activated switches and press-to-talk switches (PTT) are amongst the ones widely implemented in analogue systems. A PTT analogue switch is used here, because it provides the diver with total control over the communication process and also reduces the hardware and software complexity. When the switch is pressed, transmission mode is selected; when released, receive mode is selected. The modes of operation execute different algorithms, i.e. a speech analysis algorithm during the transmission mode and a speech synthesis algorithm during the receive mode.

The primary requirement of the transmitter is to deliver adequate power to the transducer to achieve the desired communication range of approximately 250 metres. A class E amplifier, which employs a single active component as a switch and provides approximately 90% efficiency in power transmission, is used here as the most suitable amplifier for the transmission of signals by Digital Pulse Position Modulation (DPPM).

The receiver comprises three sections: (i) preamplifier and band-pass filter; (ii) envelope detection and low-pass filter; and (iii) analogue-to-digital (A/D) converter and demodulator of the DPPM signal. Although any transducer is highly selective at around its resonance frequency, it may respond to unwanted frequencies because of the existence of spurious resonances. It is therefore necessary for the first stage low-noise preamplifier to be broadband selective at the transmit frequency. If this is not done, out-of-band ambient noise saturates the amplifier, hence limiting the detection of transmitted signals. The preamplifier is configured as a high-pass preamplifier with a cut-off frequency of 10 kHz and a gain of 20 dB. Making the preamplifier broadband selective does not provide sufficient band-pass filtering to eliminate out-of-band ambient noise; to achieve a high signal-to-noise ratio (SNR) at the receiver, the received signal must be filtered. The main specification of the 8th order active filter used here is to provide filtering within the -3dB frequencies of the transducer.

Demodulation of the DPPM signal is based on envelope detection rather than on frequency or phase measurement of the signal. The envelope of the signal is obtained with a high-precision full-wave rectifier. Next, the rectified signal is low-pass filtered to 10 kHz to smooth the waveform and to provide a baseband signal, which is then digitised. For optimum detection of the transmitted signals, the receiver bandwidth (controlled by the transducer, preamplifier and filter) must be equivalent to the bandwidth of the DPPM. To decode the DPPM baseband signals, it is necessary to apply digital signal processing techniques. Therefore, the signal must be digitised for further analysis. Before choosing an A/D converter, the sampling frequency and quantisation rate must be defined by considering input signal properties. Since the input signal bandwidth is 10 kHz, the sampling frequency must be at least twice this. Here, this is selected as 40 kHz to improve the performance of DPPM decoding.

3. SPEECH CODING

3.1 Feasibility of Speech Coding Techniques

Analogue communications systems have served the needs of divers remarkably well considering their technological simplicity. However, modern information technology requirements have introduced the need for a more robust and flexible alternative. The attraction of digitally encoded speech is obvious since it has the advantage of ease of regeneration, signaling and security, although having the disadvantage of needing extra bandwidth for transmission if it is directly applied without compression [38-46].

In principle, digital encoding of speech can be achieved by any of the well-known modulation techniques. These include pulse width modulation (PWM), pulse position modulation (PPM) and delta modulation (DM) and there are several internationally recognised speech encoding techniques listed in Table 2 whose performance is illustrated in Fig. 2. These include 64 kbit/s log-PCM (Pulse Code Modulation) and 32 kbit/s ADPCM (Adaptive Differential Pulse Code Modulation), which are classified as waveform coders [38-40]. Although they produce high quality speech and are of low complexity, they are inadequate in terms of spectrum efficiency when applied to newer bandwidth-limited communications techniques, i.e. satellite communications, digital mobile radio, private networks and especially underwater communications. These techniques are perfectly feasible for hard-wired underwater communication systems, although they do not seem to have been adopted in any presently available product and also not applied for through-water communications. There are severe problems that need to be addressed, in addition to the phenomenon of multipath interference, for underwater acoustic communication systems. One of these is related to the sampling rate requirement of speech signals. Since the speech bandwidth for telecommunication applications is about 3200 Hz (200 to 3400 Hz), then according to Nyquist's theorem the minimum sampling rate needed to recover an original analogue signal is at least twice this value. Thus, a sampling rate of 8 kHz is suitable and the corresponding sampling period is therefore 125 µs, so in the case of using waveform quantisation techniques, e.g. PCM, DM, ADPCM, this is the maximum time allowed to transmit encoded information. Owing to the limitation of the transducer used here, with a resonance frequency of 70 kHz and quality factor of Q=5, only ADPCM amongst the waveform coding techniques, with a data rate of 16 kbit/s and a bit period of 62.5 µs, may be worth considering. A necessary condition for the bit period is that it should be about Q times greater than the carrier waveform period.

It is evident from these considerations that waveform coders are not suitable for underwater acoustic voice communications, other than if a very high carrier frequency and wide bandwidth transducers are used for a very short range [47]. Since the bandwidth is severely restricted, low rate transmission is essential; hence speech signal compression is vital. For digital speech, previously not feasible, the signal compression is achieved via digital signal processing techniques. Many different techniques for compressing speech for bandwidth restricted applications have been developed [41-43]. These speech-coding techniques, classified as parametric speech coders that rely on a speech production model, compress the speech information into a few parameters that are then encoded for transmission.

Amongst the parametric coding techniques listed in Table 2, the FS1015 LPC10 speech coding (Linear Prediction Coding) standard at the bit rate of 2.4 kbit/s is used for military communications although the speech quality sounds far from natural. However, it is acceptable for secure communications. Because of in-built limitations of this voice coder, i.e. simplified model of speech production, good quality speech communication is not achievable. In order to improve speech quality and keep the bit rate lower than for waveform coders, other speech coding techniques are applicable, e.g. hybrid codecs, which use speech production models and waveform information. The FS1016 CELP (Code Excited Linear Prediction) coding technique is mainly used for secure military communications at a rate of 4.8 kbit/s, and the RPE-LTP (Regular Pulse Excited - Long Term Prediction) coding method at a rate of 13 kbit/s is implemented by GSM (Group Special Mobile) as a standard in Europe. There is also half-rate GSM using VSELP (Vector Sum Excited Linear Prediction) coding at a rate of 5.6 kbit/s [41,48-50]. These hybrid codecs produce good quality speech and have been successfully applied in mobile communications. The parametric speech coding techniques are feasible to accommodate underwater acoustic voice communications due to their low data rate and low complexity, e.g. the LPC10 vocoder. Introducing extra functions that are available in hybrid coders can also increase its performance.

3.2 Speech Production and Linear Prediction Coding

The shape of the vocal tract and the excitation mechanism determines the characteristics of a particular speech sound [51-53]. The vocal tract is considered to extend from the vocal cord to the lips and the nose, as illustrated in Fig. 3. The shape of vocal tract determines the gross frequency spectrum of a speech sound, although the effect of a sound radiating from the lips also modifies its spectrum. The resonances introduced by the vocal tract are called formants and their frequencies carry information. The formants in the spectrum are denoted F1, F2, F3, F4, F5, from lowest to highest frequency. Although there are an infinite number of formants in a given sound, usually 3-5 formants are found in the speech band.

There are three basic types of excitation in speech production, voiced, unvoiced and plosive [51-53]. Voiced sound, e.g. /i/ in *seem*, is generated by periodic opening and closing of the vocal cords and its frequency is referred to as the pitch of the speech. Unvoiced sound, e.g. /f/ in *face*, is generated by forcing air past a constriction in the vocal tract and has the characteristics of noise. Closing the vocal tract, increasing the air pressure behind it and then releasing the pressure suddenly generates plosives. The points of closure in the vocal tract determines the voiced, e.g. /b/ in *boot*, and unvoiced, e.g. /p/ in *puff*, plosive that is produced. Furthermore, sounds can be generated from a mixture of excitation mechanism, for example the /z/ in *maze* (mixed voiced and unvoiced).

From these definitions, the speech production mechanism may be modeled as shown in Fig. 4 [54], which consists of a time-varying filter and two possible excitation inputs. The excitations are impulse inputs for voiced sounds and random noise for unvoiced sounds; and the time-varying filter represents the response of the vocal tract. In parametric coding of speech signals, these two components are extracted and encoded in the transmitter, then decoded to form synthesised speech in the receiver. A Linear Prediction Coding model, as illustrated in Fig. 4, is modelled as a time-varying digital all-pole filter. This is formed by cascading a small number of two-pole resonators representing the formants and defines the envelope of the speech spectrum. The excitation and all-pole linear filter coefficients and gain parameters are estimated from a given speech sequence and quantised to achieve a transmission rate of 2.4 kbit/s [50].

3.3 Pitch Period Calculation

Pitch period is one of the essential parameters for speech synthesis and must be known in the voiced excitation state. The pitch frequency or fundamental frequency varies from 50 Hz for males to 400 Hz for females [49-53] and its accurate estimation in speech coding is needed. There are several techniques available to carry out pitch period calculations [55-63] and their performance alters with different applications, e.g. background noise affects the performance of pitch estimation. Since the speech signal produced in underwater communications is corrupted by bubble noise there should be a deviation in pitch period between clear speech and noisy speech.

The principle of pitch detection algorithms that rely on waveform similarities is to find the pitch by comparing the similarity between the original signal and its shifted version. If the shift is equal to the pitch, the two waveforms should have the greatest similarity. The auto-correlation method, which is a measure of signal agreement [59], and the average magnitude difference function (AMDF) [60], which is a measure of the disagreement, are the two most widely used techniques. The advantage of AMDF over auto-correlation is its computational simplicity, because subtraction and magnitude computations are much faster than multiply and add computations. It is applicable for real time applications of speech communications, including underwater acoustic voice communications. The AMDF is defined as:

$$AMDF(\tau) = \sum_{n=0}^{N-1} |s(n) - s(n+\tau)| \quad ; 20 \leq \tau \leq 156 \quad (1)$$

where τ is the delay. When the pitch is equal to the value of τ, the result is a minimum.

Before applying Eq. 1 to find the pitch period, the frame of the input speech signal is low-pass filtered below 900 Hz so that the second and higher formants are attenuated. Since the filtered speech signal is still composed of the first and probably the second formant frequencies, it is passed through a second-order all-zero filter to remove them and improve the performance of the pitch extractor [60-62]. Eq.1 is applied to this signal to find the minimum value of the AMDF, then the corresponding τ value, which defines the pitch period, is transmitted.

3.4 Estimation of Excitation Mode

A particular difficulty encountered in all low bit-rate voice coders, i.e. parametric coders, is the reliable and accurate measurement of the voiced/unvoiced nature of speech [51-53]. Obviously, human speech violates the simple binary voiced/unvoiced hypothesis by being a product of mixed excitation or by changing the classes over the classification interval. This is the main limitation that degrades the speech quality and intelligibility of the LPC method. Therefore, the correct voicing decision is crucial to the perceived quality and naturalness of speech and an improved voicing classification method is available for LPC applications [63]. Since the speech signal is processed in frames with LPC, a voicing decision is made for each frame. The voicing detector in the system may use the ratio of the maximum and minimum values of the AMDF function, i.e. increases if the frame is voiced. The zero crossing rate indicates the dominant spectral concentration in the speech signal. Low-band energy of the input speech signal (the most obvious and simple indicator of voiced sounds) can also be employed to improve accuracy of the voicing decision. The first two parameters are found sufficient to define a two-state excitation model given by:

$$\text{if } \begin{cases} \dfrac{max(AMDF)}{min(AMDF)} \geq 1.8 \quad \text{and} \quad \text{zero\_crossing} \leq 90 \text{ ; voiced excitation} \\ \\ \text{otherwise} \hspace{6.5cm} \text{; unvoiced excitation} \end{cases}$$

3.5 Implementation of LPC10 Algorithm

The LPC10 algorithm is implemented for underwater acoustic voice communications in order to achieve low bit-rate transmission as illustrated in Fig. 5. The speech signals are band-pass filtered at frequencies of 200 - 3400 Hz, sampled at a rate of 8 kHz, then quantised to 14-bit resolution. The digitised speech samples are stored in a buffer reserved for use by the LPC algorithm with *p=10* filter coefficients and the size of the buffer is *N=180* samples (22.5 ms), on the assumption that the speech is a stationary process. This interval is made sufficiently long to comprise at least one pitch period, i.e. approximately 20 to 160 sample intervals. In the first stage of LPC implementation, the speech signal is pre-emphasised, as defined in Eq.2, so that the energy of the high frequencies in the spectrum is increased by 6 dB per octave to compensate for the spectral effect of glottal poles [52].

$$P(z) = 1 - 0.95 z^{-1} \tag{2}$$

A windowing operation is then applied to the pre-emphasised speech signal. There are various types of windows available and their shape and length can affect the frequency representation of the speech [64]. Amongst them, the Hamming window, represented by Eq.3, is the most commonly used one, having lower spectral leakage than a rectangular window outside the frequency band of interest.

$$w(n) = \begin{cases} 0.54 - 0.46 \cos \dfrac{(2\pi m)}{N-1} & ; \ 0 \leq n \leq N-1 \\ 0 & ; \ \text{otherwise} \end{cases} \tag{3}$$

The next step is to compute the digital filter coefficients, therefore firstly the auto-correlation coefficients are calculated from the speech frame, then the reflection coefficients, $k(j)$, are estimated, as described elsewhere in detail [50].

The excitation mode and gain of the speech signal must also be defined before transmission. The excitation mode decision is made from the zero crossing and the ratio of maximum and minimum values of AMDF as described earlier. To define the gain parameter, the vocal tract model is excited by the error signal, x(n), as shown in Fig. 4 hence its energy measurement for the given speech frame can be obtained. After computing the required parameters to produce speech, they must be quantised to transmit them through an underwater acoustic communication channel.

4. ENCODING AND DECODING OF SPEECH PARAMETERS

4.1 Quantisation of Filter Coefficients

In the LPC technique, information related to a frame of the speech signal is extracted [50]. These are filter coefficients, pitch period (already coded to 6 bits), energy of the excitation input, and the excitation mode, i.e. voiced/unvoiced and 1 bit. Since a transmission rate of 2.4 kbit/s is intended the reflection coefficients and excitation gain must be quantised accordingly. There are several techniques available to achieve a low quantisation rate, such as vector quantisation [65], PARCOR (partial correlation) coefficients or reflection coefficients, k(j), [66] and line spectral frequencies [67,68]. The last two are most often used in speech communication applications. Here, the reflection coefficients, k(j), are coded to overcome the stability problem of the LPC filter, which is said to be stable if |k(j)| ≤1. The distribution plots of reflection coefficients for a long sequence of speech with a 10th order LPC filter are shown in Fig. 6 and used to define the quantisation of the reflection coefficients. Step size for uniform quantisation of the reflection coefficients, Δ, is described by

$$\Delta(j) = \frac{k(j)_{max} - k(j)_{min}}{2^{b_j} - 1}; \quad j=1,2,....,p \qquad (4)$$

where b_j is the quantisation rate for the *jth* reflection coefficient and an important parameter because of its influence in data transmission rate. The number of bits used by each coefficient results from the coefficient's relative importance in the resolution of speech. In this application, reflection coefficients, k(1) - k(4), k(5) - k(8), k(9) and k(10), are quantised at the rate of 5 bits, 4 bits, 3 bits and 2 bits, respectively.

4.2 Quantisation of Gain

As defined earlier, one of the parameters required to be transmitted in the LPC technique is the gain of the excitation signal which may be may be computed by using Eq. 5 [49,50].

$$gain = \sqrt{E_p} = \sqrt{R(0) + \sum_{j=1}^{p} a(j)R(j)} \qquad (5)$$

Only the square root of the error signal energy is computed for quantisation. The gain parameter may have a maximum value of 1024 although this is not measured during the testing of the system hence it must be considered while quantising. For this purpose, 6 bits are used. In summary, 54 bits are allocated for transmission during each 22.5 ms speech frame, as shown in Table 3. This suggests a compression ratio of 26.66 per second in the case of quantising a sample of speech at 8 bits and 46.66 per second in the case of 14 bits, which is employed here.

4.3 Decoding of Speech Parameters

On reception, the transmitted parameters must be decoded so that the speech signal is reconstructed. The decoder uses the same tables as for encoding and is accomplished by a look-up table. Once the parameters, i.e. reflection coefficients, pitch period, gain and voiced/unvoiced decision, are decoded in the receiver, they are passed to a speech synthesising stage to produce the speech signal.

5. SPEECH SIGNAL SYNTHESIS

The speech synthesising process uses the model shown in Fig. 7, which requires filter coefficients, i.e. reflection coefficient, state of excitation and gain, and pitch period parameters. Since they are already extracted during the LPC process, then transmitted by the transmitter and correctly decoded in the receiver, enough information is provided to reconstruct the speech signal. The synthesis algorithm follows from the description of the all-pole model. In the synthesiser shown in Fig. 8, a periodic impulse train for voiced speech or white noise for unvoiced speech excites ten reflection coefficients. The white noise for unvoiced excitation is generated by reading random numbers from a table with a random starting point and with a random incrementing index [69]. The excitation gain is defined for unvoiced and voiced speech in [49] as:

$$\text{Excitation Gain} = \begin{cases} \dfrac{Gain}{\sqrt{\dfrac{N}{3}}} & \text{unvoiced} \\ Gain\sqrt{\dfrac{Pitch}{N}} & \text{voiced} \end{cases}$$

A frame of delay is added in the receiver so that smoothing can be applied to the present frame. The frame-block-to-pitch-block conversion and interpolation accomplish the executive functions in the receiver. The reconstructed signal at the output of the synthesis filter is de-emphasised before application to the D/A converter. This filter, whose transfer function H(z) is the reciprocal of Eq. 2, reverses the effect of the original pre-emphasis filter [50-53]. Then the speech signal is applied to a D/A converter and low pass filter (with a cut-off frequency of 3.4 kHz) in the Analogue Interface Circuit (AIC).

The LPC programme is written in TMS320C31 DSP assembly language and stored in the memory of the prototype system. To test the algorithms, samples of the voiced part of a speech waveform, illustrated in Fig. 9, are stored in system's memory and analysed. The analysing process is done as shown in Fig. 5 and the estimated speech parameters are coded for internal transmission, i.e. modeling a noise-free channel. The received parameters are decoded accordingly and the speech signal is synthesised as shown in Fig. 7. When the LPC technique is employed for speech compression, the quality of the synthesised speech signal is significantly decreased, although speech intelligibility is maintained. Indeed, this result is expected due to the limitation of the algorithm, which only employs a binary voicing decision and two states of excitation. However, several speech coding techniques that improve the excitation input to the synthesis filter are available, e.g. CELP, RPE-LTP. Although these methods can increase the transmission rate, they are always lower than for PCM or ADPCM, so to achieve the same communication standard as mobile telephony, they must be adopted in underwater acoustic voice communications.

6. DIGITAL TRANSMISSION

6.1 Modulation Techniques

Sophisticated digital modulation methods have been adopted for radio communication systems for high data transmission. Due to the channel properties and bandwidth limitations, accomplishing the same data rate for an underwater acoustic communication channel is impractical. There are several digital modulation techniques that have been frequently used for underwater telemetry. These are Amplitude Shift Keying (ASK), Frequency Shift Keying (FSK) and Phase Shift Keying (PSK) [70-73]. The early underwater acoustic telemetry systems employed ASK and FSK for data transmission. ASK performs well when the path is straight and reverberation is low, e.g. vertical transmission. In a noisy channel its performance can be improved by introducing an error correction scheme [72] but this decreases the data rate. Since there are difficulties with ASK in a reverberant environment, most experimental underwater communication systems use some form of FSK or PSK.

FSK is normally considered as the only alternative technique for channels exhibiting rapid phase variations, such as shallow water long- and medium-range channels [73,74]. While incoherent detection eliminates the problem of carrier phase tracking, this technique is still prone to reverberation problems. FSK, which is more immune to the problems of multipath propagation, has been used for reverberant channels and performs well, for example for transmission of biomedical signals from a SCUBA diver [75].

Recent studies show that PSK achieves a better signal-to-noise ratio and is thus a frequently applied technique for underwater communications [76]. In this method, digital data is represented by phase changes, so coherent detection is required at the receiver to recover the phase information of the input signal. Therefore its implementation is more complex than the other modulation techniques. Another shortcoming of PSK is its sensitivity to multipath propagation, which causes phase and amplitude changes of the signal. To overcome coherent detection complexity, *differential* PSK (DPSK) has been developed and high rate data transmission, up to 20 kbit/s, has been achieved with a directional array [21,76]. In recent years, phase-coherent communications, 4-PSK and 8-PSK, previously not considered feasible [13, 14], have been demonstrated to be a viable way of achieving data transmission through underwater channels at rates of 4800 bit/s [77] and 10 kbit/s [78]. 4-PSK modulation has also been employed for speech transmission at 6 kbit/s [79]

To suppress ISI due to multipath propagation, non-coherent systems such as ASK and FSK employ signal formatting with guard slot intervals. They are inserted between successive pulses. These ensure that all the reverberation decays before each subsequent pulse is received [80]. The insertion of idle periods obviously results in a reduction of available data throughput. However, the advantage of these methods is that they can be easily implemented in hardware and hence are relatively simple to design. For the DPSK and M-ary PSK modulation techniques, adaptive channel equalisers have been successfully implemented to overcome the multipath propagation effect [18, 19].

6.2 Digital Pulse Position Modulation

As an alternative to the above modulation techniques, the pulse position modulation (PPM) method is also worth considering. Digital PPM has been shown to be an effective modulation format for transmitting digital information in underwater acoustic channel [81], because of its comparative simplicity in implementation and reduced sensitivity to multipath propagation. Digital information is transmitted by dividing each data frame, duration T_{symbol}, into M possible data slots, each of duration T_{slot}, and locating a transmission pulse in just one of these time slots. The mathematical definition of a DPPM pulse stream is given as

$$x(t) = \sum_{n=-\infty}^{\infty} g\left(t - nT_{symbol} - t_n\right) \tag{6}$$

where $g(t)$ is the PPM pulse shape and t_n is the random data coded into PPM such that

$$0 \leq t_n \leq (M-1)T_{slot} \tag{7}$$

To transmit quantised speech parameters, it is appropriate to select 8-slot DPPM, i.e. 3-bits per symbol. Therefore 18 symbols must be transmitted during each speech frame so that real time operation may be achieved. The symbol interval is sub-divided into eight data slots and two guard slot intervals, as shown in Fig. 10. In defining T_{symbol}, careful attention must be paid to the resonance frequency and Q of the underwater transducers; these are normally referred to as a projector-hydrophone pair but in this application each transducer acts in both transmission and reception mode. For such a transducer to be driven in its steady state, it must be excited by a sine wave of at least Q cycles duration. Another limiting factor is the rapid increase with range of the absorption loss at higher frequencies. To achieve a range of several hundred metres, as is commonly claimed for the conventional analogue systems mentioned earlier, a transducer with a 70 kHz resonance frequency and a Q factor of 5 is used. For this application T_{symbol} = 1ms, hence T_{slot} = 100 µs is defined. DPPM implementation therefore requires two clock generators. One is at the slot frequency of 10 kHz, the other at the carrier frequency determined by the hydrophone resonant frequency of 70 kHz. Two timers of the DSP are used for this purpose.

The bandwidth efficiency of PPM, i.e. R_b/B where R_b is the bit rate and B is the bandwidth, must be also considered by comparison with other digital modulation schemes. As can be seen from Table 4, by implementing DPPM, the bandwidth efficiency of the system is decreased due to an increase in the required channel bandwidth [82]. With the receiver transducer bandwidth of 25 kHz, a bit rate of 4.7 kbit/s for 8-slot DPPM can be achieved for through-water data transmission.

Since the envelope of the received signal may be used for data decoding, the binary ASK signal detection principle can be employed. Unlike for PSK there is no extra processing for phase detection and unlike for FSK there is no need for band-pass filters. Hence, in terms of system complexity also, DPPM is superior to the other techniques. However, it has several shortcomings, notably its sensitivity to multipath propagation. The destructive effect of a multipath signal on the direct path signal may therefore cause errors in data decoding. In addition, accurate synchronisation of the transmitter and receiver is required since both require accurate timing for pulse position modulation and demodulation.

6.3 Choice of Frequency Band

By contrast with the strict legal allocation of frequencies for radio transmission, there is no regulatory body to govern the allocation of frequencies for underwater communications and telemetry. In principle, any frequency can be chosen, although it is advisable to avoid the recommended emergency frequencies for manned submersible beacons (10 kHz), hazard markers (13 kHz) and diving bells (37.5 kHz) and also the carrier frequencies of communications systems used by NATO (8.0875 kHz), commercial submarines (25 kHz), the Royal Navy (40.2 kHz) and the NATO Navies (42 kHz) [83]. Frequencies of the order of 40-42 kHz have been commonly used for analogue speech communication systems between divers and between divers and the surface to give a communication range of up to 1 km depending on the transmitter's output power [84-88].

For a digital underwater voice communication system it would be desirable to set a frequency band so that the transmission of digitised speech at a rate of approximately 16 kbit/s can be achieved for a range of several hundred metres. However, it is first necessary to consider the merits of the various speech coding methods. As for satellite communication systems, there is a tendency towards the application of low bit rate speech coding and improving the decoded speech quality.

Good quality speech may be achieved for encoding techniques using bit rates of 8 kbit/s or above. In Table 5 the estimated minimum carrier frequencies for different speech coding methods and different modulation techniques are listed when the Q factor of the hydrophone is approximately 5. From Table 5, it is apparent that for digital underwater voice communications, a frequency band over 150 kHz may be used if one of the low bit rate speech coding standards is implemented, e.g. VCELP, CELP and LPC10. Although this frequency band limits the communication range, it can still be beneficial for local area communications requiring good speech quality. For the system designed and implemented in this research, a 70 kHz hydrophone was chosen, since transmission of high quality speech at a data rate of 13 kbit/s, i.e. RPE-LTP, can be accomplished even though LPC10 at 2.4 kbit/s is actually employed for simplicity.

7. TRANSMISSION AND DETECTION OF SPEECH

7.1 Encoding of Speech Parameters for Transmission

As illustrated in Fig. 11, the system includes a source encoder and a PPM encoder in the transmitter and a PPM decoder and source decoder in the receiver. The source encoder implements the speech compression algorithm and extracts appropriate speech parameters to be transmitted. The PPM-encoded speech parameters are acoustically transmitted through an underwater communication channel. Due to the nature of the communication channel, the quality of the decoded speech at the receiver is highly dependent on the performance of the detection scheme.

With the 8-slot DPPM method, the quantised speech parameters must be encoded accordingly so that transmission can be achieved. The DPPM encoder accepts as input 54 bits representing the speech parameters and groups them in 3-bit symbols stored in two 32-bit memory locations. One includes reflection coefficients $k1$ to $k6$ and $k10$, a total size of 30 bits; the other consist of $k7$ to $k9$, gain, pitch period and voiced/unvoiced decision bit, amounting to 24 bits. Once the DPPM memory locations are filled with data, the system is ready for transmission. However, since the system computes the speech parameters and transmits simultaneously, there is a difference between the transmission time (i.e. 19 ms) and the computation time (i.e. approximately 10 ms). This means that the system must be set in a waiting mode until the transmission of previous parameters is completed.

7.2 DPPM Transmission of Speech Parameters

The transmission of speech parameters is accomplished, as illustrated in Fig. 10, is accomplished by sending a synchronisation frame and 18 symbols. Symbol synchronisation is required in every digital communication system that transmits information synchronously. By the nature of DPPM, signal synchronisation is more difficult than for the more familiar digital Pulse Amplitude Modulation (PAM) systems. This is because the PPM format encodes the information by the temporal slot position of the pulse in the symbol interval. Three aspects of DPPM synchronisation are considered; these are speech frame synchronisation, DPPM symbol synchronisation and slot synchronisation, which all depend on each other [89, 90]. When the receiver is operated or if the communication is inadvertently suspended, the receiver has to be synchronised to the beginning of a speech frame in order to correctly decode the received DPPM slots.

A loss of synchronisation reduces the efficiency of the decoding process, because the inaccurate symbol synchronisation directly reduces the probability of making correct decisions. When a loss of synchronisation occurs, it may sometimes lead to successive errors before it is regained; these successive errors affect the overall performance of the system. Although the symbols are decoded properly at the receiver, they may not represent the encoded speech parameters. In some cases, the communication link may be intermittently intercepted and need to be resumed. Therefore, to synchronise the transmitter and the receiver at the beginning of each speech parameter transmission interval a synchronisation (SYNC) signal is transmitted.

From Fig. 10, it is evident that the frequency of the SYNC signal is 44.444 Hz, i.e. 1/22.5ms. To distinguish the SYNC signal from the DPPM data signal its duration must be carefully determined. After data transmission, the received signal is detected as spread out due to the response of the hydrophone and multipath propagation in the underwater acoustic channel. The DPPM data signal may occupy several successive slot intervals. This must be considered in the definition of the SYNC signal duration to be distinguished from the data slots. By taking into account all these factors, the duty cycle of the SYNC signal is defined as $t_{duty}=nT_{slot}$, where $1<n<2^k$, k is the block size, i.e. 3, and n is the number of slots, selected as 6 for this application.

7.3 Symbol Transmission

Once the system completes the transmission of the SYNC signal, the data transmission mode is selected. As described before, a timer of the DSP is used again for slot timing. When its associated interrupt request is detected, its interrupt subroutine is accessed. As soon as the subroutine is accessed, i.e. the first slot interval of the first symbol is being processed, the content of appropriate DPPM memory location is shifted to the left by 3 bits in order to form a symbol to be transmitted. Every time this interrupt subroutine is accessed (i.e. 10 times for each symbol), a 70 kHz clock signal is generated for that slot interval, the transmitter emits acoustic pulses and the DPPM data is transmitted. For a 22.5 ms speech frame represented by 54-bit quantised speech parameters, 18 symbols are required to be transmitted. The total transmission time for a 22.5 ms speech signal frame takes 19 ms for the quantised speech parameters, with 100 μs for T_{slot}. During this time the transmission, reception and decoding processes must be completed. With the speed of the DSP, the system can accomplish all this processing for two-way operation. However, owing to limitations in carrier frequency, communication range and multipath propagation, the system works in half-duplex mode. Due to the modular design of the software and hardware, the system can be modified to incorporate *two-way* operation by using a higher frequency transducer. Use of such a transducer permits implementation of Time Division Multiplexing (TDM), i.e. 11.25 ms for transmission and 11.25 ms for reception.

7.4 Detection and Decoding of Digital PPM Signal

Acoustic pulses transmitted in DPPM format through an underwater channel are detected by an omnidirectional hydrophone at the receiver, then band-pass filtered. Typical transmitted and received waveforms are shown in Fig. 12(a). Since the DPPM decoding algorithm processes the baseband signal, an incoherent detection method is used. A full-wave rectifier and low-pass filter are used to obtain the baseband signal, which is a pulse signal with an approximate pulse duration of T_{slot} (100 μs). Since the system is set to receiver mode operation, SYNC signal detection, pulse position detection and the speech synthesising algorithms are applied.

In DPPM signal detection, the limited bandwidth of the receiver affects the performance of the system by introducing a finite rise time to the received baseband signal. The slow baseband pulse build-up may result in an error in the DPPM data decoding because the necessary timing accuracy in the pulse position may not be achieved. In order to process the baseband input signal shown in Fig. 12(b), a MAX153 ADC is used, with the sampling frequency of 40 kHz and an 8-bit quantisation rate. Detection of the SYNC signal is an important feature of the receiver that is entirely based on the DPPM baseband signal. The problem is to in make a decision about the input signal because it is either the SYNC signal, the DPPM baseband signal, a multipath signal or noise. The presence of a DPPM baseband signal must be verified and the slot synchronisation must be established. In this system, this is achieved by introducing a threshold-based detection method.

If the first input sample magnitude is above the threshold level, it may not be the input DPPM baseband signal. Such a case is considered as late synchronisation of the baseband signal. If the first sample is the DPPM signal, the system does not perform any decoding process until speech frame synchronisation is established. Here, we assume that the first sample is taken from the SYNC signal. If this is not the closest sample to the rising edge and the system is allowed to continue the DPPM baseband signal decoding, it is likely that a high error rate will occur. This is simply because of the accurate timing relation between the SYNC signal and the DPPM signal. Therefore, it is important that the first sample magnitude is lower than the threshold level so that slot synchronisation is established. This phenomenon is called as early synchronisation of the baseband signal. The source for this sample is assumed to be either from noise or a multipath signal. The receiver continuously samples the input baseband signal and compares its magnitude with the fixed threshold value. As soon as a sample whose magnitude is higher than that of the threshold level is detected, the slot synchronisation is verified and the receiver is enabled to search the SYNC signal.

When slot synchronisation is achieved with a fixed threshold level, a search is made for the speech frame synchronisation signal. The baseband signal is sampled at 40 kHz, i.e. 4 samples per slot, and quantised at 8-bit resolution. Then the energy of each slot, $E[slot]$, is calculated as:

$$E[slot] = \sum_{k=1}^{4} x[k]^2 \qquad 0 \leq slot \leq 7 \qquad (8)$$

where k is the sample number, $x[k]$ is the digitised baseband signal. Six of the ten slot intervals are reserved for the transmission of the SYNC signal. The energies of these six slots should be higher than that during the non-pulse intervals. Moreover, because of the fixed threshold level it is possible to define the expected minimum slot energy value. After the slot synchronisation is completed, the energy of each successive slot is calculated and compared to the pre-defined energy level. When six successive slot energies are found to be equal to or greater than the threshold energy level, this is the condition that speech frame synchronisation has been established.

If speech frame synchronisation is not achieved, the DPPM decoding process is not implemented. Moreover, because the rising edge of the SYNC signal is taken as a reference in the demodulation of the DPPM signal, the sharper the edge and the higher the sampling frequency the more accurate the pulse position detection achievable. Since the DPPM demodulator works on the position estimation principle, a timing error significantly increases the bit error rate. Once accuracy in the SYNC signal detection is achieved, the DPPM slot data decoding can be easily implemented.

During trials the transmitter, unit 1, and the receiver, unit 2, were closely positioned so that the magnitude of the multipath signal was minimised, although it is more significant in the SYNC signal detection. The receiver demodulates the DPPM baseband signal, without reference to the threshold level, which is only used during synchronisation. Here, we assume that the transmitter and the receiver are synchronised. Baseband signal energy for each of the 8 slot intervals, is calculated by means of Eq. 8 and the results are stored in eight memory locations to be decoded during the guard interval.

The maximum energy, which indicates the possible pulse position, is accepted as the position of the transmitted digital data. The same processing is done for the subsequent 17 symbol intervals. While the DPPM decoding is in process, the data are further decoded to form the quantised speech parameters. By implementing such a decoding process, the transmitted speech parameters are successfully recovered and used in the speech synthesising algorithm. During DPPM decoding, the multipath signal energy is assumed to be relatively small compared with the direct path signal energy. Provided that perfect synchronisation is achieved, the decoded speech parameters should be free of error.

8. DISCUSSION

The aim of this chapter has been to describe the design of a novel digital underwater acoustic voice communication system. Speech encoding and decoding algorithms are implemented to achieve modulation and demodulation of speech data. When the two systems were tested off-line, where an almost perfect square wave baseband signal is obtained and multipath signal and channel noise are absent, the problems discussed above are not encountered. This confirms that both systems operate according to the design criteria. During through-water testing, one of the two unit was configured as a transmitter, the other as a receiver.

The encoded speech parameters were successfully transmitted through water and decoded at the receiver, then the speech signals were synthesised, as illustrated in Fig. 13. As expected, the speech quality at the receiver was synthetic. With the current speech-coding algorithm, achieving a good quality speech signal is a difficult task that requires modification of the algorithms. During the testing of these two systems, occasional interrupts in the synthesised speech signal occurred, which resulted in the generation of noise for a period of 22.5 ms. The interrupts were caused by a SYNC signal detection error. During the synthesis of speech signals, the contents of the speech parameters are modified. When there is an error in the SYNC signal detection, the synthesising algorithm still uses these incorrect speech parameters. This is the reason why noise is detected. A possible solution is to use previous speech parameters. For this particular voice communication system, no bit error rate measurements are performed due to the hardware limitations. However, such an analysis is essential to illustrate the performance of the system and to verify the efficiency of the modulation and demodulation tasks.

9. REFERENCES

[1] Woodward, B. "Underwater Telephony: Past, Present and Future," *Colloque De Physique*, Colloque C2, No.2, pp. C2_591-C2_594(1990).

[2] Baume, D., Godden, D. and Hipwell, J. " Improving diver communications," *Underwater Systems Design*, pp.21-23 (December 1979/January 1980).

[3] McIntosh, W., "Underwater communications: An oil company viewpoint," *J. Soc. Underwater Technology*, Vol.11, No.3, pp.14-26 (1985).

[4] Hicks, R.J. and Virr, L.E. "Underwater communications-a review," *Int. Conf. Divetech'81: The Way Ahead in Diving Technology*, Society for Underwater Technology, London, (1981).

[5] Bud, A.M.G., Holmes, J., White, I., Datham, R.J. and Kramer, J. "Improvements in diver communications" *Report OTI 87 500*, Dept. Energy, H.M. Stationary Office (1986).

[6] Anderson, V.C., "Acoustic communication is better than none," *IEEE Spectrum*, pp.63-68 (1970).

[7] Berktay, H.O., Gazey, B. and Teer C.A. "Underwater communication past, present and future," *J. Sound Vib.* Vol.7, Part 1, pp.62- 70 (1968).

[8] Hollien, H., Coleman, R.F., and Rothman, H.B. "Evaluation of diver communication systems by a diver-to-diver technique," *IEEE Trans. Communication Tech.*, Vol. COM-19, No.4, pp.403-409 (1971).

[9] Webb, H.J., and Webb, J.R., "An underwater audio communicator," *IEEE Trans. Audio Electroacouctics*, Vol.AU-14, No.3, pp.127-135 (1968).

[10] Urick, R.J. *Principles of Underwater Sound*, McGraw-Hill, New York (1975).

[11] Burdick, W.S. *Underwater Acoustic System Analysis*, Prentice-Hall, New Jersey (1984).

[12] Coates, R.F.W. *Underwater Acoustic Systems*, Macmillan, London (1990).

[13] Catipovic, J.A. "Performance limitations in underwater acoustic telemetry," *IEEE J. Oceanic Eng.*, Vol.OE-15, No.3, pp.205-216 (1990).

[14] Baggeroer, A.B. " Acoustic telemetry - an overview," *IEEE J. Oceanic Eng.*, Vol.OE-9, No.4, pp.229-234 (1984).

[15] Jourdain, D., "Acoustical propagation analysis in shallow water," *Proc. 2nd European Conf. Underwater Acoustics*, Vol.II, pp.781-786 (1994).

[16] Owen, R.H., Smith, B.V. and Coates, R.F.W. "An experimental study of rough surface scattering and its effects on communication coherence," in *Proc. Oceans'94*, pp.III.483-III.488 (1994).

[17] Geller, B., Brossier, J.M., and Capellano, V., "Equalizer for high data rate transmission in underwater communications," in *Proc. Oceans'94*, pp.I.302-I.306 (1994).

[18] Geller, B., Capellano, V., Brossier, J.M., Essebbar, A. and Jourdain, G. "Equalizer for video rate transmission in multipath underwater communications," *IEEE J.Oceanic Eng*, Vol.21, No. 2 (1996).

[19] Proakis, J.G. "Adaptive equalization techniques for acoustic telemetry channels," *IEEE J. Oceanic Eng.*, Vol.OE-16, No.1, pp.21-31 (1991).

[20] Stojanovic, M. "Recent advances in high-speed underwater acoustic communications," *IEEE J.Oceanic Eng*, Vol.21, No. 2, pp. 125-136 (1996).

[21] Tarbit, P.S.D, Howe, G.S., Hinton, O.R., Adams, A.E. and Sharif, B.S. "Development of a real-time adaptive equalizer for a high-rate underwater acoustic data communications link," in *Proc. Oceans'94*, pp.I.307-I.312 (1994).

[22] Bessios, A.G. and Caimi, F.M. "Multipath compensation for underwater acoustic communication," in *Proc. Oceans'94*, pp.I.317-I.322 (1994).

[23] Galvin, R. and Coates, R.F.W. "Analysis and performance of an underwater acoustic communications system and comparison with a stochastic model," in *Proc. Oceans'94*, pp.III.478-III.482 (1994).

[24] Billon, D. and Quellec, B. "Performance of high data rate acoustic underwater communication systems using adaptive beamforming and equalizing," in *Proc. Oceans'94*, pp.III.507-III.512 (1994).

[25] Henderson, G.B., Tweedy, A., Howe, G.S., Hinton, O., and Adams, A.E., "Investigation of adaptive beamformer performance and experimental verification of applications in high data rate digital underwater communications," in *Proc. Oceans'94*, pp.I.296-I.301 (1994).

[26] Mackie, R.D.L., and Smith, N. "Diver communications and helium speech unscrambling," *Underwater System Design*, pp.19-21 (June/July 1983).

[27] Mendel, L.L., Hamill, B.W., Crepeau, L.J. and Fallon, E. " Speech intelligibility assessment in a helium environment," *J. Acoust. Soc. Am.*, Vol. 97, No. 1, pp. 628-636 (1995).

[28] Fant, G. and Sonesson, B. "Speech at high ambient air pressure," in *Speech Transmission Lab. Quart. Prog. and Status Report*, STL-QPSR No. 2/1964, pp.9-21 (1964).

[29] Fant, G. and Lindqvist, J. " II. Studies related to diver's speech. A: Pressure and gas mixture effects on diver's speech," in *Speech Transmission Lab. Quart. Prog. and Status Report* STL-QPSR No. 2/1968, pp.7-17 (1968).

[30] Richards, M.A. "Helium speech enhancement using the short-time Fourier transform," *IEEE Trans. on Acoust. Speech and Signal. Process.* Vol.ASSP-30, No.6, pp.841-853 (1982).

[31] Jack, M.A. and Duncan, G. "The helium speech effect and electronic techniques for enhancing intelligibility in a helium-oxygen environment," *The Radio and Electronic Engineer*, Vol.52, No.5, pp.211-223 (1982).

[32] Mackie, R.D.L., and Smith, N. "Diver communications and helium speech unscrambling," *Underwater System Design*, pp.19-21 (June/July 1983).

[33] Beet, S.W. and Goodyear, C.C. "Helium speech processor using linear prediction," *Electronics Letters*, Vol.19, No.11, pp.408-410 (1983).

[34] Brandt, J.F. and Hollien, H. " Underwater hearing thresholds in man as a function of water depth," *J. Acoust. Soc. Am.*, Vol.46, No.4, pp.893-894 (1969).

[35] Brandt, J.F. and Hollien, H. " Underwater hearing thresholds in man," *J. Acoust. Soc. Am.*, Vol.42, No.5, pp.966-971 (1967).

[36] Wong, W.T.K., Mack, R.M., Cheetham, B.M.G., and Sun, X.Q. "Low rate speech coding for telecommunications," *BT Technol. J.* Vol.14, No.1, pp.28-43 (1996).

[37] Barret, P.A., Voelcker, R.M., and Lewis, A.V., "Speech transmission over mobile radio channels," *BT Tech. J.*, Vol.14, No.1, pp.45-55 (1996).

[38] Koo, B. and Gibson, J.D. " Experimental comparison of all-pole, all-zero and pole-zero predictors for ADPCM speech coding," *IEEE Trans. Commun.*, Vol.Com-34, No.3, pp.285-290 (1986).

[39] Bonnet, M., Macchi, O. and Saidane, M.J. " Theoretical analysis of the ADPCM CCITT algorithm," *IEEE Trans. Commun.*, Vol.38, No.6, pp.847-858 (1990).

[40] Sherif, M.H., Bowker, D.O., Bertocci, G., Orford, B.A., and Mariano, G.A. "Overview and performance of CCITT/ANSI embedded ADPCM algorithms," *IEEE Trans. Commun.*, Vol.41, No.2, pp.391-399 (1993).

[41] Spanias, S.A. "Speech coding: A tutorial review," *Proc. IEEE*, Vol.82, No. 10, pp.1541-1582 (1994).

[42] Jayant, S.N. " Coding speech at low bit rates," *IEEE Spectrum*, pp.58-63 (1986).

[43] Carmody, J. and Rothweiler, J. "Speech coding at 800 and 400 bps," *Electrical Commun.* pp.260-265 (1986).

[44] Benvenuto, N., Bertocci, G., Daumer, W.R. and Sparrell, D.K. " The 32-kb/s ADPCM coding standard," *AT&T Tech. J.*, Vol.65, pp.12-19 (1986).

[45] Lafuente, L.M. "Adaptive differential pulse code modulation for low bit rate transmission of speech signals," *Electrical Commun.*, Vol.58, No.2, pp.225-229. (1983).

[46] Tremain, T.E. " The government standard linear predictive coding," *Speech Technology*, Vol.1, pp 40-49 (1982).

[47] Woodward, B. and Sari, H. "Digital underwater acoustic voice communications," *IEEE J.Oceanic Eng*, Vol.21, No. 2, pp. 181-192 (1996).

[48] Samuel, D.S. and Ruth, A.D, *Signal Processing Algorithms*, Prentice-Hall, New Jersey, (1989).

[49] Markel, J.D. and Gray, A.H, *Linear Prediction of Speech*, Springer-Verlag, Berlin (1982).

[50] Makhoul, J. "Linear Prediction: A tutorial review," *Proc. IEEE*, Vol.63, No.4, pp.561-580 (1975).

[51] Rabiner, L.R. and Schafer, R.W, *Digital Processing of Speech Signals*, Prentice-Hall, New Jersey (1978).

[52] Deller, Jr. J.R., Proakis, J.G., and Hansen, J.H.L, *Discrete-Time Processing of Speech Signals*, Macmillan Publishing Comp., New York (1993).

[53] Saito, S. and Nakata, K, *Fundamentals of Speech Signal Processing*, Academic Press, Tokyo (1985).

[54] Flanagan, J.L., Ishizaka, K., and Shipley, K.L. " Synthesis of speech from a dynamic model of the vocal tract," *The Bell System Tech. J.*, pp.485-506 (1975).

[55] Krubsack, D.A. and Niederjohn, R.J. " An autocorrelation pitch detector and voicing decision with confidence measures developed for noise-corrupted speech," *IEEE Trans. Signal Proc.*, Vol.39, No.2, pp.319-329 (1991).

[56] Tsakalos, N. and Zigouris, E. " An investigation of failures and comparison of correlation measurement pitch trackers and pre-processing filters," *Int. J. Electronics*, Vol.75, No.2, pp.269-283 (1993).

[57] Gold, B. and Rabiner, L.R. " Parallel processing techniques for estimating pitch periods of speech in the time domain," *J. Acoust. Soc. Am.*, Vol.46, pp.442-448 (1969).

[58] Sukkar, R.A., Locicero, J.L. and Picone, J.W. " Design and implementation of a robust pitch detector based on a parallel processing technique," *IEEE J. Select. Areas Commun.*, Vol.6. No.2., pp.441-450(1988).

[59] Rabiner,L.R. "On the use of autocorrelation Analysis for pitch detection," *IEEE Trans. Acoust., Speech, and Sig. Proc.*, Vol.ASSP-25, No.1, pp.24-33 (1977)

[60] Ross, M.J. " Average magnitude difference function pitch extractor," *IEEE Trans. Acoust., Speech, Sig. Proc.*, Vol.ASSP-22, pp.353-362 (1974).

[61] Tsakalos, N., and Zigouris, E. " Threshold based magnitude difference function pitch determination algorithm," *Int. J. Electronics*, Vol.71, No.1, pp.13-28 (1991).

[62] Markel, J.D. "The SIFT algorithm for fundamental frequency estimation," *IEEE Trans. Audio Electroacoust.*, Vol.AU-20, pp.367-377 (1972).

[63] Campell, P.J., and Tremain, E.T. "Voiced/unvoiced classification of speech with applications to the U.S. Government LPC-10E algorithm," *Int. Conf. Acoustic, Speech and Signal Proc.*, pp.473-476 (1986).

[64] Ifeachor, C.E., and Jervis, B.W., *Digital Signal Processing- A Practical Approach*, Addison-Wesley, Wokingham (1993).

[65] Buzo, A., Gray, A.H., Gray, M.R. and Markel, J.D. "Speech Coding Based upon Vector Quantization," *IEEE Trans. Acoustic, Speech, Sig. Proc.*,Vol. ASSP-28, pp.562-574 (1980).

[66] Gray, A.H, Gray, R.M., and Markel, J.D. "Comparison of optimal quantization of speech reflection coefficients," *IEEE Trans. Acoustic, Speech, Sig. Proc.*,Vol. ASSP-25, No.1, pp.9-21 (1977).

[67] Itakura, F. "Line spectrum representation of linear predictive coefficients of speech signals," *J. Acoust. Soc. Am.*, Vol.57, pp.535 (A) (1975).

[68] Lepschy, A., Mian, G.A., and Viaro, U. "A note on line spectral frequencies," *IEEE Trans. Acoustic, Speech, Sig. Proc.*, Vol.36, No.8, pp.1355-1357 (1988).

[69] "Details to assist in implementation of Federal Standard 1016 CELP," National Communications System, Tech. Bulletin, 92-1 (1992).

[70] Brock, D.C., Bateman, S.C. and Woodward, B. " Underwater acoustic transmission of low-rate digital data," *Ultrasonics*, Vol.24, pp.183-188 (1986).

[71] Andrews, R.S. and Turner, L.F. "On the performance of underwater data transmission system using amplitude shift keying techniques," *IEEE Trans. on Sonic and Ultrasonics*, Vol.SU-23, pp.64-71 (1976).

[72] Dawoud, M.M., Halawani, T.U. and Abdul-jauwad, S.H. "Experimental realisation of ASK underwater digital acoustic communications system using error correcting codes," *Int. J. Electronics*, Vol.72, No.2, pp.183-196 (1992).

[73] El-Hennawey, M.S. and Shehadah, W.H. "Non-coherent FSK receiver for underwater communications," *Int. J. Electronics*, Vol.79, No.3, pp.265-280 (1995).

[74] Catipovic, J., Baggeroer, A.B., Von Der Heydt, K. and Koelsch, D. "Design and performance analysis of a digital acoustic telemetry system for the short-range underwater channel," *IEEE J.Oceanic Eng*, Vol. OE-9, pp.252-252 (1984).

[75] Woodward, B. and Bateman, S.C. "Diver monitoring by ultrasonic digital data telemetry," *Med. Eng. Phys.* Vol. 16, pp.278-286 (1994).

[76] Thompson, D., Neasham, J., Sharif, B.S., Hinton, O.R., Adams, A.E., "Performance of coherent PSK receivers using adaptive combining and beamforming for long range acoustic telemetry," *3rd European Conf. Underwater Acoustics*, pp.747-752 (1996).

[77] Falahati, A., Bateman, S.C. and Woodward, B. " Underwater acoustic channel models for 4800bps QPSK signals," *IEEE J. Oceanic Eng.*, Vol.OE-16, No.1, pp.12-20 (1991).

[78] Stojanovic, M., Catipovic, J.A. and Proakis, J.G. "Phase-coherent digital communication for underwater acoustic channels," *IEEE J. Oceanic Eng.*, Vol.19, No.1, pp.100-111 (1994).

[79] Goalic, A., Labat, J., Trubuil, J., Saoudi, S. and Riouaten, D. "Toward a digital acoustic underwater phone," in *Proc. Oceans'94*, pp.III.489-III.494 (1994).

[80] Habib Istepanian, R.Sh., *Use of Microcontrollers for Diver Monitoring by Underwater Acoustic Biotelemetry in Multipath Environments*, Ph.D. Thesis, Loughborough University (1994).

[81] Riter, S. and Boatrigth, P.A. "Design considerations for a pulse position modulation underwater acoustic communication system", *Digest IEEE Conf. Eng. Oceanic Environment*, pp.21-24 (1970).

[82] Proakis, J.G., and Salehi, M., *Communication Systems Engineering*, Prentice-Hall, New Jersey (1994).

[83] Sear, J.K. "Standardisation of Sonar Communications," *Int. Conf. Divetech'81: The Way Ahead in Diving Technology*, Society for Underwater Technology, London, (1981).

[84] Mulcahy, M., "A through-water diver communication system," *Sea Technology*, Vol.20, No.8, pp.27-29 (1979)

[85] Gazey, B.K. and Morris, J.C., "An underwater acoustic telephone for free-swimming divers," *Electronic Eng.*, pp.364-368 (1964).

[86] Peck, M.J. " Wireless underwater communications past, present, and future," *Sea Technology*, pp.61-65 (1992).

[87] Overfield, T. " Modern Through-Water Diver Communications," *Underwater Systems Design*, pp.8-13 (March/April 1988).

[88] Clark, A. " Diver communications- The case for Single Sideband," *Underwater Systems Design*, pp.16-18 (January 1989).

[89] Ling, G. and Cagliardi, R.M. " Slot synchronization in optical PPM communications," *IEEE Trans. Commun.*, Vol.COM-34, No.12, pp.1202-1208 (1986).

[90] Georgehiades, C.N. "Optimum joint slot and symbol synchronization for optical PPM channel," *IEEE Trans. Commun.*, Vol.COM-35, No.6, pp.518-527 (1987).

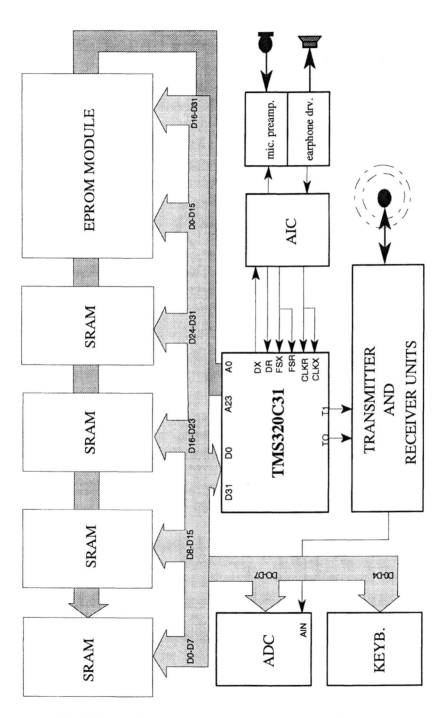

Fig. 1 Digital underwater voice communications system architecture

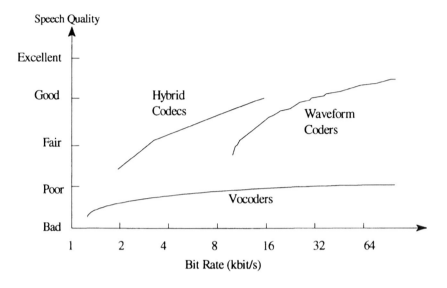

Fig. 2 Comparison of speech coding techniques [52]

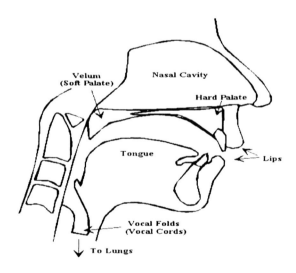

Fig. 3 Human speech production mechanism [49]

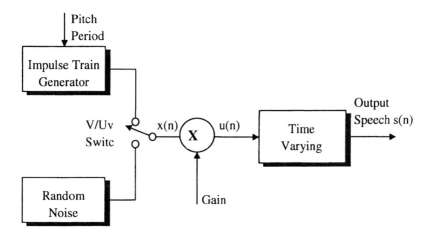

Fig. 4 Simplified source filter model of speech production

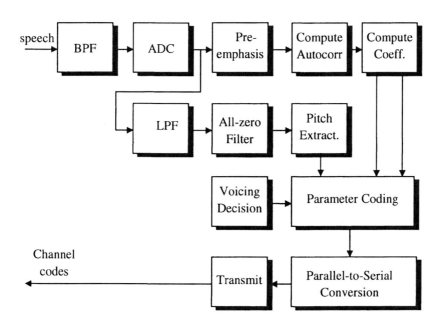

Fig. 5 Linear predictive speech coder

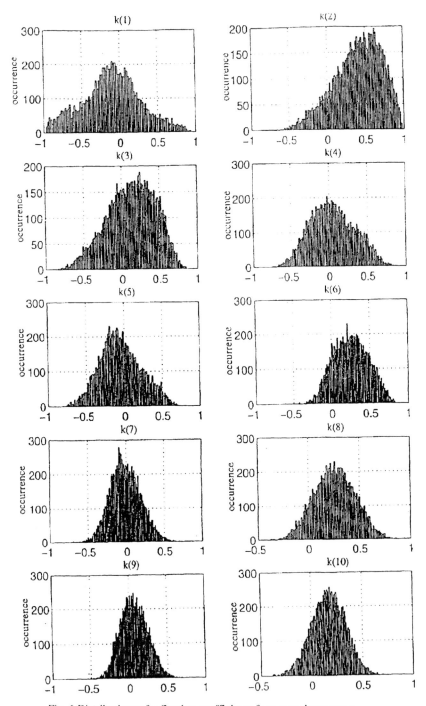

Fig. 6 Distributions of reflection coefficients for a speech sequence

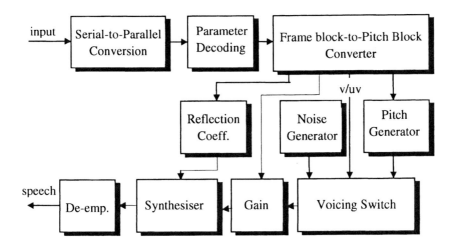

Fig. 7 Linear predictive code synthesiser

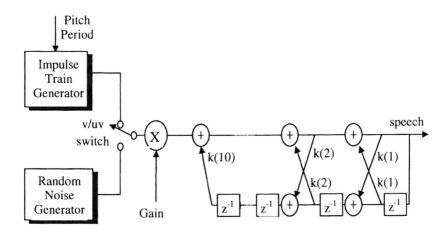

Fig. 8 Synthesiser

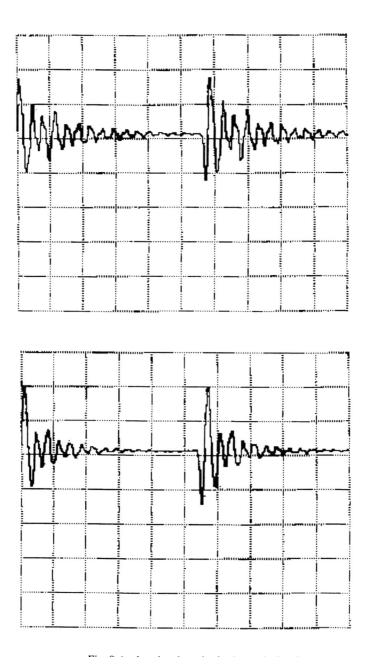

Fig. 9 Analysed and synthesised speech signal

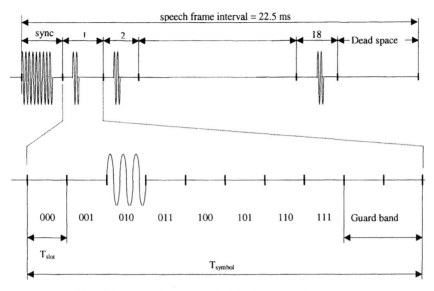

Fig. 10(a) Transmission of digital data in DPPM format

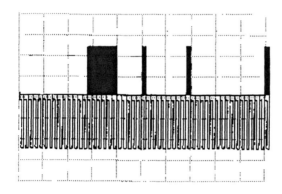

Fig. 10(b) Transmission of SYNC signal and octal data in DPPM format

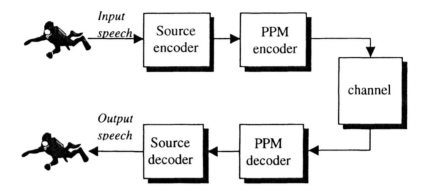

Fig. 11 Underwater communication system

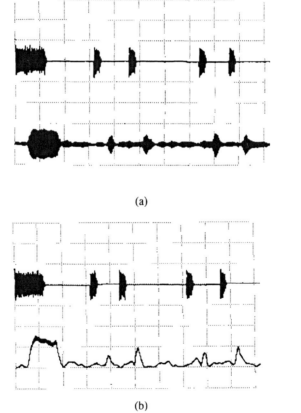

(a)

(b)

Fig. 12 Transmitted and received DPPM waveforms (a) output of bandpass filter (b) output of the envelope detector (Ch1:10V/div; Ch2:2V/div; time/div=0.5ms).

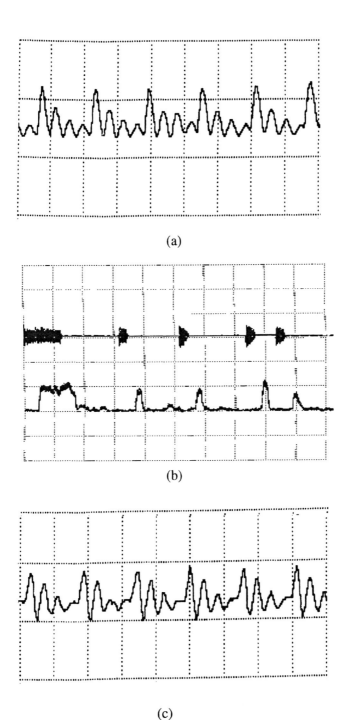

Fig. 13 Speech communications using DPPM: (a) input speech signal, (b) transmission of coded speech parameters, (c) synthesised speech signal

Resonant frequency (f_o)	69.8 kHz
Q factor	4.1
Low cut-off frequency	61 kHz
High cut-off frequency	78 kHz
Susceptance in water (B_o)	3.117×10^{-3} S
Conductance in water (G_o)	3.615×10^{-3} S
Equivalent resistance (R_s)	276.6 Ω
Static capacitance (C_o)	7.107 pF
Bandwidth :	18 kHz

Table 1 Properties of the transducer at resonance

Standard	Year	Coding Type	Bit Rate(kbit/s)	Class
ITU-G.711	1972	PCM	64	Waveform Coding
ITU-G.721	1984	ADPCM	32	
ITU-G.727	1991	VBR-ADPCM	16,24,32,40	
US DoD FS1015	1984	LPC-10	2.4	Parametric Coding
GSM Full-rate IS54	1989	RPE - LTP	13	
US DoD FS1016	1990	CELP	4.8	
IS-96	1991	Qualc. CELP	1,2,4,and 8	
GSM Half-Rate	1994	VSELP	5.6	

Table 2 Speech coding standards

Parameter	Bits
gain	6
pitch	6
k(1)	5
k(2)	5
k(3)	5
k(4)	5
k(5)	4
k(6)	4
k(7)	4
k(8)	4
k(9)	3
k(10)	2
voice/unvoiced	1
total	54

Table 3 Bit allocations for LPC parameters

Modulation Techniques	M	Bandwidth Efficiency	M	Bandwidth Efficiency
M-ary ASK	2	0.5	8	1.5
M-ary PSK	2	0.5	8	1.5
M-ary FSK	2	0.25	8	0.1875
M-slot DPPM	2	0.25	8	0.1875

Table 4 Bandwidth efficiency of digital modulation techniques

speech encoding	bit rate kbit/s	ASK f_0(kHz)	FSK f_0-f_1(kHz)	PSK f_0(kHz)	QPSK f_0(kHz)	DPPM f_0(kHz)
PCM	64	320	320-448	320	160	1066.6
ADPCM	16	80	80-112	80	40	266.6
LD-CELP	16	80	80-112	80	40	266.6
RPE-LTP	13	65	65-91	65	32.5	216.66
VCELP	8	40	40-56	40	20	133.33
CELP	4.8	24	24-33.6	24	12	80
LPC10	2.4	12	12-16.8	12	6	40

Table 5 Recommended carrier frequencies for digital underwater voice communications for different speech coding and modulation methods (8-slot DPPM is assumed)

5

Applications of Neural Networks in Underwater Acoustic Signal Processing

Zheng Zhaoning
Department of Radio Engineering, Southeast University, Nanjing, Jiangsu 210018, P.R.China.

1. INTRODUCTION

It is well known that underwater acoustic channels are particularly complicated and difficult for typical signal processing procedures due to the time-varying, homogeneous volume, rough boundaries, and the abundance of interference noise sources in these channels. Most traditional methods of statistical signal processing employ simplified assumptions (linear, stationary, Gaussian) for the sake of mathematical tractability that inevitably lead to inadequate performance.

In this context, the rapidly developing neural networks provide a viable alternative for underwater acoustic signal processing and communication techniques. These methodologies can indeed solve problems that are difficult for traditional signal processing methods. In recent years, neural networks have been successfully applied to underwater acoustical signal processing applications. Examples include sonar targets detection and classification, optimal array processing, noise removal for undersea communication, ocean modeling, intelligent control for autonomous underwater vehicles.

The main aim of this chapter is to present some aspects of the application of neural networks in underwater acoustic signal processing with emphasis on both methodology and implementation issues. In section 2, we present some of fundamentals of neural networks. Section 3 is devoted to the implementation problems of neural networks in general. In this section two digital implementation architectures: Multi-DSP systems and VLSI array processors are also presented. Illustrative examples together with their use in implementing specific neural networks (Hopfield and RBF networks) are discussed in detail. In section 4, the DOAs estimation problem is formulated, and the way of using neural networks in solving the DOAs estimation problem is explained. The parallel Tank-Hopfield networks is then presented as an illustrative example, which offers an immediate means to realize the most advanced DOAs estimation techniques in real-time and can be implemented immediately with analog VLSI technology.

2. FUNDAMENTALS OF NEURAL NETWORKS

2.1 Basic Concepts

Neural networks (NNs), or more precisely artificial neural networks (ANNs), have been studied for more than 50 years and have gone through the first upsurge, the period of stagnation, and the new resurgence in the last two decades. Interest in neural networks is prompted by the hope of achieving the performance of the human nervous system, which can easily and efficiently solve problems that are exceedingly difficult for conventional techniques. Broadly speaking, neural networks are composed of a group of simple nonlinear computational elements (neurons) with extensive and weight-changeable connections (synapses) among them. The neurons are usually divided into layers and arranged in structures that are more or less imitative of biological neural systems. Neurons can work in continuous-time or discrete-time and be implemented with analog or digital VLSI. Neural networks may be viewed from different perspectives as follows.

1. <u>Neural networks are simplified models of human brain</u>. They can calculate and reason. They can learn from examples and encode the knowledge learned in patterns of synaptic weights. So they are capable of associative memory when the patterns are recalled and compared with input data which are also called input patterns. They can adapt to the environment by adjusting their free parameters and even changing their structure or their problem solving strategy. They are also good at tackling fuzzy or low precise information. Consequently, they possess obvious advantages over traditional techniques in the areas of pattern recognition and adaptive control.

2. <u>Neural networks are massively parallel and distributed processors</u>. They can provide much higher computation rate than that of a von Neumann computer in which a program of instructions is performed sequentially. Thus they have the potential to implement complicated computation in real-time. For example, the computation of matrix and its inversion or pseudo inversion, the computation required for solving linear or quadric optimization problems. They are inherently fault tolerant because they consist of a large number of neurons each with primarily local connection. Damage to a few neurons or their synaptic links need not impair overall performance significantly.

3. <u>Neural networks are generally nonlinear systems</u> due to the fact that in most cases each neuron in a neural network is a nonlinear processing unit. Moreover, some neural networks are universal approximators in the sense that when trained with supervision they can approximate any continuous nonlinear input-output mapping to any desired accuracy. Therefore, neural networks are particularly suited for modeling and solving nonlinear problems in real world.

4. <u>Neural networks are nonparametric processors</u>. They make no assumption about the underlying probability distribution. In many practical applications,

such as the target detection and location in an undersea environment, it is impossible to characterize the statistics of the received waveforms fully and precisely. The conventional approaches based on assumptions of Gaussianity, linearity and stationarity often lead to serious performance deterioration. In principle, the nonparametric approach may provide more robustness and capability for detecting signals generated by nonlinear and non-Gaussian processes.

5. <u>Neural networks are dissipative dynamical systems</u>. That is, the time evolution of system states can be described by deterministic rules that may be, among others, differential equations or iterative equations. The methods and results of the dynamical system theory can be used to understand, analyze and design neural networks. For example, we may employ Lyapunov's direct method to judge the stability of a neural network. However, because of the intractability due to the nonlinearity and highly complex structure, the dynamical behavior of most neural networks are not well understood and hard to analyze. The mathematics of neural networks is still in its infancy.

A neural network can be entirely specified by three basic components: the neuron (node) model or its input/output equation, the network topology, and the training or learning rule. In the sequel, we will discuss them briefly.

2.2 Neuron Model

Neurons are usually classified into two categories: static and dynamic.

Static neuron is memoryless. Its output is only a function of the current input and has nothing to do with past or future inputs or outputs. The input/output equation of a static neuron can be expressed in the general form

$$v(t) = f(u(t))$$

$$u(t) = g(x_1(t), \cdots, x_n(t); w_1, \cdots, w_m; c_1, \cdots, c_k)$$

1

where $v(t)$ is the output state of the neuron, $u(t)$ is the internal state, $x_1(t), \cdots, x_n(t)$ are current inputs, $w_1, \cdots, w_m$ are connection weights, $c_1, \cdots, c_k$ are some parameters.

The simplest and most widely used static neuron model is Perceptron, where

$$u(t) = \sum_{i=0}^{n} w_i x_i(t), \quad x_0(t) \equiv 1, \; w_0 \equiv \theta \qquad 2$$

and $f(\cdot)$ is a hard limiter, or a sigmoid function, or a linear function. Another example is the neuron with an input/output relation as

$$u(t) = -\frac{1}{2\sigma^2} \sum_{i=1}^{n} [x_i(t) - w_i]^2 \qquad 3$$

$$f(\cdot) = \exp(\cdot)$$

The functional relation between $v(t)$ and $\{x_i(t)\}$ is called radial basis function (RBF).

Dynamic input/output equations, on the other hand, are of memory and usually have the form of differential or difference equations. As an example, the neuron used in the continuous-time Hopfield Network is given in Figure 1. Its input / output equation can be expressed in the form

$$C \frac{du(t)}{dt} = -\frac{u(t)}{R} + \sum_{i=1}^{n} w_i x_i(t) - \theta \qquad 4$$

$$v(t) = f(u(t))$$

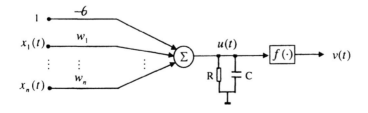

Figure 1- Dynamic Neuron Used in Continuous-Time Hopfield Network

2.3 Network Topologies

Network topology refers to the geometrical pattern in which neurons are arranged and interconnected. There are many terms to describe such a pattern: single-layer, multilayer, feedforward, feedback, and so on. Two most popular multilayer feedforward neural networks are the Multilayer Perceptron (MLP) and the Radial Basis Function Network (RBF), shown in Figure 2.

In the MLP, each node is a Perceptron and is fully connected in the sense that for every node in layer k, its inputs come from all the nodes in layer $k-1$ and its output feeds into every node in layer $k+1$. The MLP can be used in implementation of arbitrary logic functions, in pattern classification and functional approximation. Three layers are enough for most applications. In the RBF, the output layer consists of linear Perceptrons and neurons in the hidden layer have nonlinearity of radial basis function. It can be shown that both MLP and RBF are capable of forming an arbitrararily close approximation to any continuous nonlinear mapping by properly selecting weights and using sufficient number of hidden units. In this sense they are universal approximators.

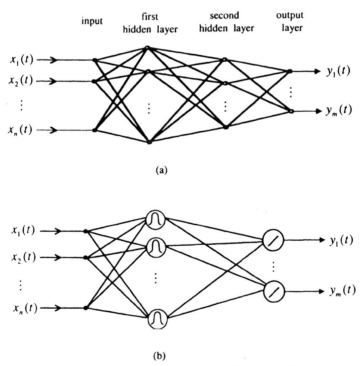

Figure 2- Multilayer Feedforward Neural Network. (a) Three-Layer Perceptron (MLP). (b) Radial Basis Function Network (RBF).

Network topology with feedback connections is also extensively used. This class of networks is commonly referred to as the recurrent neural network because they are inherently recursive. Two important examples are given in Figure 3. The first one in Figure 3(a) is the continuous-time Hopfield network. It is a single-layer network where each node takes the form as presented in Figure 1 and every node output $v_i(t)$ is connected to any other node input, and also in many cases to its own input. Furthermore, each node may receive external input b_i. The second one in Figure 3(b) is discrete-time Hopfield network, in which each node is a Perceptron with nonlinearity of hard limiter and its inputs come from outputs of the network.

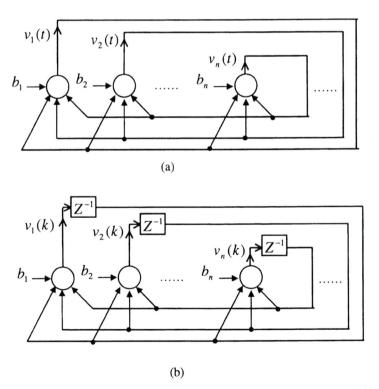

Figure 3 - Neural Network with Feedback Connections. (a) Continuous-Time Hopfield Network. (b) Dicrete-Time Hopfield Network.

2.4 Training Rule

When neuron model and network topology are given, training or learning refers to the procedure, with a preset rule, of finding all the synaptic connection weights with which the network works optimally according to a criterion. There are two kinds of training: supervised and unsupervised.

In the case of supervised training, a set of training pair $\{\mathbf{x}_p, \mathbf{d}_p\}_{p=1}^{P}$ is provided, where $\mathbf{x}_p$ is an n dimensional input signal, $\mathbf{d}_p$ is the corresponding m dimensional desired output or teacher signal. During the training, input $\mathbf{x}_p$ is applied, the network will successively modify its connection weights with a training rule in an attempt to make its output $\mathbf{y}_p$ equal to or approximate optimally to the desired $\mathbf{d}_p$. The whole training procedure takes turns utilizing all the training pairs once or many rounds until a criterion is satisfied. For instance, the total error $\sum_{p=1}^{P}(\mathbf{y}_p - \mathbf{d}_p)^T(\mathbf{y}_p - \mathbf{d}_p)$ is below a permitted value or the weights are near their convergence values.

In the case of unsupervised training, no teacher signal is provided. The modification of connection weights only depends on input patterns and network states. During training, the network can automatically extract features from input patterns and accordingly form clusters or vector quantization among them. Such a training scheme is also called self-organization.

2.5 Some Comments on Neural Networks

Neural networks possess many distinguished features and advantages. They have been successfully used to solve problems in various areas with performance obviously better than traditional techniques. However, it never means neural networks will inevitably replace conventional approaches, because not only parallel computing is not the natural or optimal approach to all problems, but also neural networks have their own disadvantages and limitations. Firstly, due to the nonlinearity and highly complex structure, a perfect and firm theoretical foundation of neural networks is far from being established. We lack analytical means for understanding the stability and convergence properties of various training algorithms. We are also in want of comparative and quantitative performance measures. To some extent, a neural network is like a black box. We can not exactly know the details of its dynamical behaviors. Basically, various neural network models and training algorithm now available are all based on experience and *ad hoc* strategies. Secondly, in order to demonstrate their ultimate capabilities, neural networks must be implemented in parallel computing hardware devices. For accommodating large numbers of inputs and extensively distributing and processing the data, the hardware is typically widespread and expensive. Thirdly, to ensure robust performance, the neural network must access to fairly large training sets of data which should be representative of the classes to be recognized. In many applications, comprehensive and enough training sets are very difficult to obtain at the design stage. The neural network must collect data and learn from them at worksite. However, the training algorithms employed by most neural networks are so time-consuming that online adaptability/training performance in changing environments is prevented.

It is universally believed that ongoing works in the development of hardware devices, entirely new types of neural networks, and mathematical foundations, will overcome most of the disadvantages just mentioned. On the other hand, the traditional computer techniques are mature and successful in numerical computation and logical operation, and have accumulated a great wealth of problem-solving software sources. We ought to sensibly use achievements in conventional computer techniques to enhance the ability of neural networks. It needs to solve the interface problem between these two kinds of techniques.

3. IMPLEMENTATION ISSUES OF NEURAL NETWORKS

3.1 A Brief Review

In order to implement a neural network means to carry out all computations involved in its training and test modes. This is a difficult task because neural networks usually possess huge amount of computation, massive interconnection and complicated training rules. The simplest approach utilizes a single general-purpose computer to perform the task. However, due to serial operations in traditional Von Neumann Computer, the computation rate is too low to ensure this approach being useful in most practical applications. A computer specially designed for implementing neural networks with high efficiency is called a Neural Network Computer (NNC). There are four levels in the design procedure of a NNC (in descending order): application, neural network model, NNC architecture, and physical processing element (PE). We must first define the application tasks of a NNC. For example, it is supposed to be used in sonar target detection or/and recognition or/and location. Or it is supposed to be of general- or special-purpose. For performing a specific application task, we have a variety of neural network models to choose from. When a neural network model has been selected, there are still many choices for the NNC architecture. The most popular structure of NNC is nowadays a host computer plus a group of PEs. In this context, the architecture problem deals with the way in which all PEs are combined and work together. For example, the data exchange between any two PEs is executed through the bus or through direct local communication, all PEs work synchronously or asynchronously. As for PEs, the various choices are digital or analog, semiconductor chip or optical device, simple VLSI circuit or complicated DSP, etc. In a sense, the NNC design procedure is just to make all the above choices as good as possible to meet the application requirements and real world constraints.

In this section we are mainly concerned with the NNC architecture problem when using DSPs or digital VLSI circuits as PEs. Here one of the central issues is *efficiency*, i.e. how to increase the amount of computation within a fixed time interval, or, in other words, how to raise the level of parallelism. Recently, the two prevalent schemes are

• Multi-DSP systems, in which each PE is a complicated programmable DSP with a local memory, and works in parallel and asynchronously with others through the shared main memory and data set. A DSP typically implements several hundreds of neurons. Because the number of DSPs is much less than that of neurons, multi-DSP systems are known as virtual implementation and being of coarse-grained parallelism. A NNC with the architecture of a host plus multi-DSP system belongs to MIMD (Multiple Instruction stream Multiple Data stream) computer. Multi-DSP systems can be divided into two categories. In the first, all data exchanges among DSPs are executed only through the bus. In the second, direct communications among DSPs are also available. The use of DSPs such as TMS320C40s or Transputer T800s, for example, falls in the second case.

• VLSI array processors, in which each PE is an application-specific or reconfigurable VLSI circuit with a local memory, and can only perform simple arithmetic and logic operation. All PEs are regularly interconnected to form an array. Then the regular communication, control and pipeline techniques are introduced into the array. The parallelism in VLSI array processors can be either 'coarse-grained' or 'fine-grained'. It depends on whether the VLSI array possesses a smaller number of relative powerful circuit units or a larger number of very elementary circuit units. A NNC with the architecture of a host plus VLSI array processor belongs to SIMD (Single Instruction stream Multiple Data stream) computer. VLSI array processors can also be further classified, according to the pattern how PEs are driven to work. Two subclasses are of clock-driven synchronous pattern (systolic array) and data-driven asynchronous pattern (wavefront array).

Another central issue in the NNC architecture problem is *flexibility*. It concerns how to raise the programmable ability. The balance between efficiency and flexibility is the most fundamental tradeoff in the design of NNC. For the implementation schemes mentioned above, at one extreme are application-specific VLSI circuit array processors, which contain dedicated circuitry optimized to the application task at hand, thus achieving the highest efficiency at the lowest silicon cost. However, they suffer the drawback of being only useful for the one task for which they were designed. At the other extreme are the multi-DSP systems, where the programmable DSP implements a *fixed* set of arithmetic and control operations that can be organized and *sequenced* to perform any arbitrary computation. The flexibility of DSP comes at the cost of efficiency. When the native operations of DSP are not well suited to the given task, or when the massive amounts of parallelism can be exploited, multi-DSP systems are inefficient and offer relative poor performance. In the middle between these two extremes are the reconfigurable VLSI array processors, where the VLSI circuits are implemented with *programmable logic* which is a flexible hardware capable of being structured to fit the nature organization and data flow of a computation. Today's available device for reconfigurable computing is the Field-Programmable Gate Array (FPGA), in which each PE is a logical unit with an ability ranging from powerful to very elementary, and the interconnections among logical units are programmable. FPGAs can work in highly parallel and deeply pipeline fashion as any VLSI array processor does. They also allow a direct and nature approach to reconfigue the logical units interconnected to implement operations required by the application. They can even track the change of the computational requirement imposed by the changing processing task or data. As a newly emerging technique, FPGAs have not been widely adopted nowadays because of their high power consumption, low clock speed, long reconfiguration time, and the lack of a general high-level software programming tool. However, along with the advances of fabrication technology and the realization of plans for a billion transistors on a chip, more serious consideration will be paid to the reconfigurable VLSI array processors for their use in the implementation of neural networks.

In the following, we will illustrate some of these implementation schemes.

3.2 Implementation by Multi-DSP Systems

As mentioned above, there are two kinds of multi-DSP system based NNCs. The difference lies in whether the data exchanges among DSPs are executed only through the bus or the direct communication among DSPs are also available. Some details about these methods will be given here for illustration. We will also show how to use these to implement a specific neural network. In this context, two main considerations are

- Searching for the most efficient algorithm. A neural network model can usually be calculated by different algorithms with different convergence rates and computational complexities. Choosing an algorithm best suited to the given NNC architecture may be critical in some application case.

- Decomposing the whole computational task into a set of subtasks, each for a DSP. In a good decomposition, the computation load of each DSP should be balanced and the communication overhead should be as small as possible. Decomposition obviously depends on the algorithm and the NNC architecture in use.

Multi-DSP system based NNCs where the data exchanges among DSPs are executed only through the bus.

Many commercially available NNCs fall in this category, such as Mark III from TRW Company, Odyssey from Texas Instruments, etc. They are all designed for general purpose to support the research of neural networks by speeding up their programming and operation. For example, at most 15 DSPs are employed in Mark III, where each DSP consists of a MC68020 plus a floating point processor MC68881, and all DSPs are connected to a host computer VAX through a VME bus.

For the sake of illustration, we present here a relative simple architecture of small size and being easy to realize. It is called Neuro C (Zhang et al. 1995),as shown in Figure 4. There are five identical DSP modules in Neuro C, one of which is used as host, others to perform the computational task. The DSP selected is microprocessor i486. Many other kinds of DSP, such as TMS32000 series, Transputer series, i860/960 series, may work as well. The Transmission BUS supports the communication among all DSP modules. It can perform the peer to peer and selected broadcast communication. The selected communication means that the broadcast targets may be any combination of DSP modules 1, 2, 3, 4 and host. The local memories in all DSP modules are arranged in a single addressing memory space. Such an arrangement is called a distributed-shared-memory system. Each DSP module can access data in local memory at a lower cost, and data in other DSP module's memory without difficulty. So advantages of both shared and

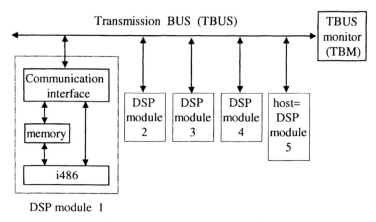

Figure 4 - The Architecture of Neuro C

distributed memory system are gained. The programming is as simple as in the usual single memory space. The software of Neuro C is also well developed, which includes the debug environment and tools, the communication library, the monitor/loader of each DSP module, the C++ -like programming language and its compiler.

Now we illustrate how to use Neuro C to implement a discrete-time Hopfield network of order 512 (see Figure 3(b) with n=512). The iterative equation is

$$\mathbf{v}(k+1) = \text{sgn} \ [\mathbf{W}\mathbf{v}(k) + \mathbf{b} \ ] \qquad\qquad 5$$

$$v_i(k+1) = \text{sgn}[\sum_{j=1}^{512} w_{ij} v_j(k) + b_i], \ i = 1, \cdots, 512$$

where $\text{sgn}[x] = 1$, when $x > 0$; $= 0$, otherwise. The most part of computational overhead lies in calculating the matrix-vector multiplication. The whole computation can be decomposed into four subtasks by partitioning the matrices involved

$$\begin{bmatrix} \mathbf{v}_1(k+1) \\ \cdots \\ \mathbf{v}_2(k+1) \end{bmatrix} = \begin{bmatrix} \mathbf{W}_{11} & \vdots & \mathbf{W}_{12} \\ \cdots & \cdots & \cdots \\ \mathbf{W}_{21} & \vdots & \mathbf{W}_{22} \end{bmatrix} \begin{bmatrix} \mathbf{v}_1(k) \\ \cdots \\ \mathbf{v}_2(k) \end{bmatrix} + \begin{bmatrix} \mathbf{b}_1 \\ \cdots \\ \mathbf{b}_2 \end{bmatrix} = \begin{bmatrix} \mathbf{W}_{11}\mathbf{v}_1(k) + \mathbf{W}_{12}\mathbf{v}_2(k) + \mathbf{b}_1 \\ \cdots \\ \mathbf{W}_{21}\mathbf{v}_1(k) + \mathbf{W}_{22}\mathbf{v}_2(k) + \mathbf{b}_2 \end{bmatrix}$$

where $\mathbf{v}_i$ and $\mathbf{b}_i$ are 256×1 vectors, $\mathbf{W}_{ij}$ is a 256×256 matrix, $i, j = 1$ or 2. The DSP module 1 performs the computation of $\mathbf{W}_{11}\mathbf{v}_1(k)$, the DSP module 2 performs the computation of $\mathbf{W}_{12}\mathbf{v}_2(k)$, etc. The partial results are all sent to the host, where addition and sign operation are executed. Then the final results $\mathbf{v}_1(k+1)$, $\mathbf{v}_2(k+1)$ are sent back to relevant local memories to start the next iteration.

Multi-DSP system based NNCs where the direct data communications among DSPs are also available.

This type of NNCs is usually of less general-purpose, so we will only give an illustrative example here (DiZitti et al. 1989). The architecture is shown in Figure 5. It is a 2×2 array of DSPs connected to a master and a host. Here Transputers are used for the DSPs and the master. However, the more powerful TMS320C40 or C80 are also good choices. Each DSP in the architecture can execute its own program to perform the

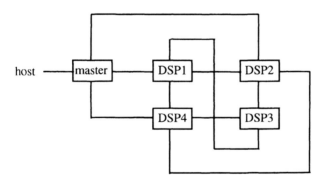

Figure 5 - Structure of a 2×2 Array of Transputers Connected to a Master and a Host

computation assigned. At the same time, it can communicate with other DSPs and the master by asynchronous handshaking through four two-way links. DSP3 is an exception, which communicates with master via DSP1. The master controls the whole computational procedure in the array by receiving, synthesizing and sending relevant information from/to DSPs.

We now again illustrate the use of this type of NNCs for implementing a Hopfield network. The first thing we have to do is searching for the most efficient algorithm. According to Figure 1, 3(a), and equation 4, the i-th neuron state $u_i(t)$ of the original continuous-time Hopfield network is expressed in the form

$$\frac{du_i(t)}{dt} = -A_i u_i(t) + \sum_{\substack{j=1 \\ j \neq i}}^{n} w_{ij} f(u_j(t)) + b_i, \quad i = 1, \cdots, n \qquad 6$$

where $f(\cdot)$ is a sigmoid function, $A_i^{-1} = C_i R_i$ is the time constant of the i-th neuron (see Figure 1), b_i is the external constant excitation. After discretization by Euler method, or replacing $du_i(t)/dt$ by $[u_i(t+\Delta t) - u_i(t)]/\Delta t$, it becomes

$$u_i(t+\Delta t) = (1 - A_i \Delta t) u_i(t)$$
$$+ \sum_{\substack{j=1 \\ j \neq i}}^{n} w_{ij} f(u_j(t)) \Delta t + b_i \Delta t, \quad i = 1, \cdots, n \qquad 7$$

Equation 7 is called *improved* discrete-time Hopfield network model as compared to equation 5, because it preserves both the continuous and nonlinear properties of the neuron as much as possible and the advantage of simple iteration.

However, the original Hopfield network model in equation 6 can also be discretized by the multistep integration method that leads to the following Gauss-Seidel relaxation algorithm (DiZitti et al. 1989),

$$u_i(t+\Delta t) = (1 - A_i \Delta t) u_i(t) + \sum_{j=1}^{i-1} w_{ij} f(u_j(t+\Delta t))\Delta t$$

$$+ \sum_{j=i+1}^{n} w_{ij} f(u_j(t))\Delta t + b_i \Delta t, \quad i = 1, \cdots, n \qquad 8$$

We then have three algorithms at hand. They are equations 5, 7 and 8. The best choice among them depends on the comparison of their performance. Equation 5 can be easily ruled out. After having compared the rate of convergence, the computational complexity and the dependence of solution on initial conditions between algorithms 7 and 8 both theoretically and in computer simulation, it reveals that the Gauss-Seidel algorithm 8 converges faster than the improved discrete-time algorithm 7 when applying them to solve the optimization problems of A/D conversion and the traveling salesman. A gain factor up to 100 has been observed for the A/D problem. Moreover, the Gauss-Seidel algorithm is more robust in that the problem of being trapped in local minima is less serious.

Once the Gauss-Seidel algorithm has been chosen, the next step for decomposition is simple and straightforward. All neurons are divided into four groups of equal number $n/4$. The p-th DSP performs the computational task for the p-th group, in which the i-th neuron's output iterates as:

$$u_{i,p}(t+\Delta t) = (1 - A_{i,p}\Delta t)u_{i,p}(t) + \sum_{j=1}^{i-1} w_{ij}^{(p)} f(u_{j,p}(t+\Delta t))\Delta t$$

$$+ \sum_{j=i+1}^{n/4} w_{ij}^{(p)} f(u_{j,p}(t))\Delta t + b_{i,p}\Delta t$$

$$+ \sum_{\substack{q=1 \\ q \neq p}}^{4} \sum_{j=1}^{n/4} w_{ij}^{(q)} f(u_{i,q}(t))\Delta t \qquad\qquad 9$$

where $p,q = 1,2,3,4$ are DSP or group index, $i, j = 1, \cdots, n/4$ are neuron index within each group. We note that only the *local* data are needed for calculating the first four terms in equation 9, and only the *previous* data are used to compute the last term.

3.3 Implementation by VLSI Array Processors (a Systolic Array Example)

In this implementation framework, an array of application specific VLSI circuits performs the neural network computation under the control of a host. For the purpose of illustration, we will present a systolic array example. A systolic array is a network of digital VLSI signal processing elements (PEs) which have both distributed computation and distributed memory, and rhythmically compute and pass data through the network. It provides an effective means of combining a large number of PEs with only nearest neighboring interconnections and has the important features of modularity, regularity, minimum control overhead, highly pipelined and synchronized multiprocessing. The systolic array can efficiently implement advanced matrix operations such as matrix-matrix multiplication, matrix factorization, matrix inversion, singular value decomposition, and least squares (LS) approximate solution of system of linear equations. In the following we discuss how a systolic array can be used to implement a RBF network. The key idea is that we can transform the training procedure of a RBF network into a problem of finding the LS solution of a system of linear equations. Consequently, we need only to choose a proper algorithm that makes systolic array implementation feasible and effective. Figure 6 shows a RBF network that we want to implement with a systolic array. It consists of :

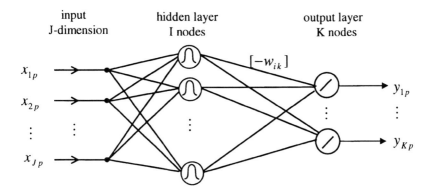

Figure 6 - A RBF Network to be Implemented with a Systolic

a J-dimensional input, a hidden layer with I nodes, and an output layer with K nodes. Let $(1 \times J)$ input vector $\mathbf{x}_p = [x_{1p}, x_{2p}, \cdots, x_{Jp}]$, $(1 \times K)$ output vector $\mathbf{y}_p = [y_{1p}, y_{2p}, \cdots, y_{Kp}]$. the output of the i-th node in the hidden layer can be expressed as (see equation 3)

$$\exp\{-\frac{1}{2\sigma^2}\sum_{j=1}^{J}(x_{jp}-x_{ji}^{(c)})^2\} \triangleq \phi(\|\mathbf{x}_p - \mathbf{x}_i^{(c)}\|) \qquad 10$$

where $\mathbf{x}_i^{(c)} = [x_{1i}^{(c)}, x_{2i}^{(c)}, \cdots, x_{Ji}^{(c)}]$ is the preset $(1 \times J)$ center vector of the radial basis function in the i-th node. So we have

$$y_{kp} = -\sum_{i=1}^{I}\phi(\|\mathbf{x}_p - \mathbf{x}_i^{(c)}\|)w_{ik} \qquad 11$$

where $-w_{ik}$ is the ik-th element of the $(I \times K)$ weight matrix $-\mathbf{W} = [-w_{ik}]$.

Suppose a set of training pair $\{\mathbf{x}_p, \mathbf{d}_p\}_{p=1}^{P}$ is given. The training procedure searches for a weight matrix $-\mathbf{W}$ that minimizes the total squared error

$$\sum_{p=1}^{P}(\mathbf{d}_p - \mathbf{y}_p)(\mathbf{d}_p - \mathbf{y}_p)^T = \sum_{p=1}^{P}\sum_{k=1}^{K}(d_{kp} - y_{kp})^2$$

$$= \sum_{p=1}^{P}\sum_{k=1}^{K}[d_{kp} + \sum_{i=1}^{I}\phi(\|\mathbf{x}_p - \mathbf{x}_i^{(c)}\|)w_{ik}]^2$$

When $P > I$, it is equivalent to find a LS approximate solution of the following overdetermined system of linear equations

$$d_{kp} + \sum_{i=1}^{I}\phi(\|\mathbf{x}_p - \mathbf{x}_i^{(c)}\|)w_{ik} = 0, \quad k = 1, \cdots, K; p = 1, \cdots, P \qquad 12$$

with KP

$$\mathbf{\Phi W} + \mathbf{D} = \mathbf{O} \qquad 13$$

where $\mathbf{\Phi} \triangleq [\phi_{pi}] = [\phi(\|\mathbf{x}_p - \mathbf{x}_i^{(c)}\|)]$ is a $(P \times I)$ matrix, $\mathbf{D} \triangleq [\mathbf{d}_1^T, \cdots, \mathbf{d}_P^T]^T$ is a $(P \times K)$ matrix, $\mathbf{d}_p = [d_{1p}, \cdots, d_{Kp}]$ is a $(1 \times K)$ row vector, $\mathbf{O}$ is $(P \times K)$ matrix with zero elements.

The LS solution can be computed via the **QR** decomposition of $\mathbf{\Phi}$. That is to find a $(P \times P)$ matrix $\mathbf{Q}$ which columns are orthonormal such that

$$\mathbf{Q\Phi} = \begin{bmatrix} \mathbf{R} \\ \cdots \\ \mathbf{O} \end{bmatrix}, \quad \mathbf{QD} = \begin{bmatrix} \mathbf{U} \\ \cdots \\ \mathbf{V} \end{bmatrix} \qquad 14$$

where **R** is a $(I \times I)$ upper triangular matrix, **U** is a $(I \times K)$ matrix, **O** is a $(I \times K)$ matrix with zero elements. We then only need to solve

$$\mathbf{R}\mathbf{W} + \mathbf{U} = \mathbf{O} \qquad 15$$

However, direct **QR** decomposition requires global data communication. For applying the systolic array to implement a RBF network, we have to divide the whole **QR** decomposition into operations that only utilize local data. In this context, the most popular method is to calculate **QΦ** by means of a series of simple matrix multiplication

$$\mathbf{Q}\mathbf{\Phi} = \mathbf{Q}(P)\mathbf{Q}(P-1)\cdots\mathbf{Q}(2)\mathbf{Q}(1)\mathbf{\Phi}$$

$$\mathbf{Q}(p) = \mathbf{Q}(p+1, p)\mathbf{Q}(p+2, p)\cdots\mathbf{Q}(P, p)$$

The matrix $\mathbf{Q}(q, p)$ is called *Givens rotation operator*, which is quite easy to calculate and requires only local data neighboring to the (q, p) element (Kung 1985).

Broomhead et al. (1989) have proposed a systolic array scheme for implementing a RBF Network, as shown in Figure 7. There are four kinds of PEs in the array. Individual PE functions are given in Figure 8. Each PE in the array passes data to or receives data from its nearest neighbor in the directions indicated by the arrows and is synchronized to perform within a single clock cycle. In order to ensure that each PE processes proper data at proper time, the data-flow must be duly arranged. The arrangement includes the data skewness in the input vectors **x** and **d**, a single clock cycle delay between the neighboring PEs in the edge GI (represented by a black dot), and a waiting time of $(J + 1 + I)$ cycles for the first training data vector $\mathbf{d}_1$ with respect to the corresponding $\mathbf{x}_1$.

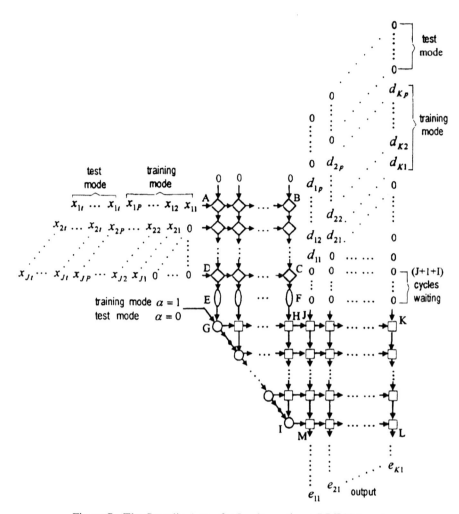

Figure 7 - The Systolic Array for Implementing a RBF Network

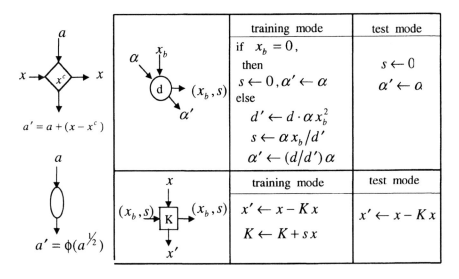

Figure 8 - Four Kinds of PEs and their Functions

The whole network comprises three connected blocks. The ((J+1)×I) rectangle block ABFE produces a sequence of I-dimensional vector

$$\{[\phi(\|\mathbf{x}_p - \mathbf{x}_1^{(c)}\|), \cdots, \phi(\|\mathbf{x}_p - \mathbf{x}_I^{(c)}\|)]\}_{p=1}^{P}$$

in a skew fashion at the edge EF when skew input vectors $\{\mathbf{x}_p\}_{p=1}^{P}$ are sequentially applied at the edge AD. The (I×I) triangle block GHI is the Gentleman-Kung array. In training mode, a parameter α is initially set to one for the first PE at the edge GI. Each PE in the edge GI computes the parameters of Givens rotations and passes them to the right, where the internal PEs perform the rotation by utilizing the data entering from above. The (I×K) rectangle block JKLM consists only of internal PEs. In training mode, it performs rotations on the training data vector $\mathbf{d}_p$.

During training, the content in local memory of each PE (K's and d's) within the blocks GHI and JKLM is updated in such a way that at the end of training the scaled versions of **R** and **U** in equation 15 are contained in the blocks GHI and JKLM, respectively. Furthermore, the output vector $\mathbf{e}_p = [e_{1p},\cdots,e_{Kp}]$ emerged at the lower edge ML represents the residual error vector,

$$e_{kp} = d_{kp} - y_{kp} = d_{kp} + \sum_{i=1}^{I} \phi(\|\mathbf{x}_p - \mathbf{x}_i^{(c)}\|) w_{ik} \qquad 16$$

The LS optimal weight matrix **W** is implicit in the array memory. It is not necessary to solve equation 15. In fact, for operating in the test mode, all we have to do is setting $\alpha = 0$ for the first PE at the edge GI and replacing **d** with **0** at the edge JK. In such a case, the update of all local memory in blocks GHI and JKLM stops and the output in equation 16 becomes

$$e_{kt} = -\sum_{i=1}^{I} \phi(\|\mathbf{x}_t - \mathbf{x}_i^{(c)}\|) w_{ik} \qquad 17$$

which is just the output of a RBF network after training.

The complete network operates in a highly parallel and deeply pipelined fashion when all PEs work synchronously. However, when system synchronization becomes a critical problem, the wavefront array based on asynchronous data-driven computing provides an alternate choice (Kung 1985).

4. APPLICATIONS TO EIGENVECTOR-BASED DOA ESTIMATION

The localization of radiating sources by a sensor array is one of the central problems in underwater acoustics, and has attracted considerable interest recently. The simplest problem in this context is the estimation of the directions-of-arrival (DOAs) of multiple narrow-band plane waves with the same known center frequency. The radiating sources are assumed in the far field. The array can be of arbitrary geometry or regular shape, e.g. uniform linear array (ULA) or uniform circular array. We will only discuss two-dimensional DOAs (azimuthal directions) estimation here for the sake of clarity. However, the approach can be extended straightforwardly to the three-dimensional (azimuth and elevation) and multiple parameters (DOAs, range, depth, frequency) estimation case. It can also be applied to a wide variety of problems including accurate detection and estimation of sinusoids in noise, improvement of communication system performance in severe multipath environment, etc.

4.1 Problem Formulation and Assumptions

We consider a two-dimensional array of M sensors and arbitrary geometry. Suppose there are d sinusoidal point sources in the far field, all have the same frequency ω. At the origin of coordinates, the complex envelope of the plane wave signal from the i-th source is denoted by $s_i(t)$ and its direction of arrival (DOA) relative to the horizontal axis

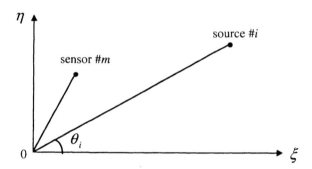

Figure 9 - Two-Dimensional Array geometry

by θ_i, as shown in Figure 9. The output complex envelope of the m-th sensor generated by the i-th source can be expressed in the form

$$x_{m,i}(t) = g_m(\theta_i) e^{-j\frac{\omega}{c}[\xi_m \cos\theta_i + \eta_m \sin\theta_i]} s_i(t) \hat{=} a_m(\theta_i) s_i(t) \qquad 18$$

where c is sound velocity, ξ_m and η_m are the spatial coordinates of the m-th sensor, $g_m(\theta_i)$ is the directional response of the m-th sensor at frequency ω and direction θ_i. The whole output of the m-th sensor can be obtained by summing up contributions of all sources,

$$x_m(t) = \sum_{i=1}^{d} a_m(\theta_i) s_i(t) \qquad 19$$

Let $\mathbf{x}(t) \triangleq [x_1(t), \cdots, x_M(t)]^T$, $\mathbf{a}(\theta_i) \triangleq [a_1(\theta_i), \cdots, a_M(\theta_i)]^T$. The array output vector takes the form

$$\mathbf{x}(t) = \sum_{i=1}^{d} \mathbf{a}(\theta_i) s_i(t) \qquad 20$$

By defining

$$\boldsymbol{\theta} \triangleq [\theta_1, \cdots, \theta_d]^T \qquad 21$$

$$\mathbf{A}(\boldsymbol{\theta}) \triangleq [\mathbf{a}(\theta_1), \cdots, \mathbf{a}(\theta_d)]_{M \times d} \qquad 22$$

$$\mathbf{s}(t) \triangleq [s_1(t), \cdots, s_d(t)]^T \qquad 23$$

we have the model commonly used in array processing when an additional noise $\mathbf{n}(t)$ exists:

$$\mathbf{x}(t) = \mathbf{A}(\boldsymbol{\theta}) \mathbf{s}(t) + \mathbf{n}(t) \qquad 24$$

The sample spatial covariance matrix of $\mathbf{x}(t)$ is given by

$$\hat{\mathbf{R}} = \frac{1}{N}\sum_{i=1}^{N}\mathbf{x}(t_i)\mathbf{x}^H(t_i) \qquad 25$$

Where $t_i = iT$, T is the temporal sampling period, N is the total number of observation data, H means the conjugate transpose. $\hat{\mathbf{R}}$ can be viewed as a proper estimate of the covariance matrix $\mathbf{R}$,

$$\mathbf{R} \triangleq E[\mathbf{x}(t)\mathbf{x}^H(t)]$$
$$= \mathbf{A}(\boldsymbol{\theta})E\{\mathbf{s}(t)\mathbf{s}^H(t)\}\mathbf{A}^H(\boldsymbol{\theta}) + E\{\mathbf{n}(t)\mathbf{n}^H(t)\} \qquad 26$$

Almost all array processors now available are based on second-order statistics of data, i.e. employ the information provided by $\mathbf{R}$. To make the problem analytically tractable, a few reasonable assumptions are adopted as follows:

1. $\mathbf{n}(t)$ is a stationary, temporal white and zero-mean Gaussian process. For simplicity we also require $\mathbf{n}(t)$ to be spatially white, i.e.

$$E\{\mathbf{n}(t)\mathbf{n}^H(t)\} = \sigma^2 \mathbf{I}$$

The temporal and spatial white assumptions are not serious restrictions in that if the temporal and spatial noise covariance are known or have been estimated, the noise can be pre-whitened through simple procedures.

2. The number of sources is smaller than that of sensors, i.e. $d<M$. Let source covariance matrix be

$$E[\mathbf{s}(t)\mathbf{s}^H(t)] = \mathbf{S}_{d \times d}$$

and the rank of $\mathbf{S}$ be d'. We have $d'=d$ (full rank) in the case of all sources being non-coherent, $d'<d$ (rank deficiency) in the case where two or more coherent sources are present. Two signals are regarded as coherent if they are different only in an amplitude factor and a time delay. Coherent signals usually originate from the multipath propagation effect caused by sea bottom or surface reflection.

3. $\mathbf{A}(\boldsymbol{\theta})$ is of full rank.

4. The number of sources d is known a priori or through some estimation procedures.

The problem of central interest herein is to estimate the source directions $\{\theta_1,\cdots,\theta_d\}$ when given a finite data set $\{x(t)\}$ observed over a period of time $\{t_1,\cdots,t_N\}$. We are only concerned with DOAs estimation approaches based on second-order statistics. Basically, there are two types of such approaches: <u>conventional approaches</u> that make use of the information provided by $\mathbf{R}$ simply and directly, and <u>advanced or eigenvector-based approaches</u> that utilize the inherent geometrical structure information of $\mathbf{R}$ in more ingenious fashions. The famous eigenvector-based approaches are MUSIC, Min-Norm, Root-MUSIC, WSF, DML, AM, ESPRIT, etc.

4.2 The Use of Neural Networks in implementing Eigenvector-Based DOAs Estimation

The eigenvector-based methods exhibit much better resolution power than that of conventional methods, but their implementation in real time is much difficult because of very expensive computation. The bottleneck lies in the performing of EVD, or the calculation of a certain projective matrix. For example, we must calculate

$$\mathbf{P}_{A(\theta)} = \mathbf{A}(\theta)[\mathbf{A}^H(\theta)\mathbf{A}(\theta)]^{-1}\mathbf{A}^H(\theta)$$

for WSF, DML or its alternative maximization version AM, and calculate

$$\hat{\mathbf{P}}_s^\perp = \hat{\mathbf{U}}_n \hat{\mathbf{U}}_n^H$$

for MUSIC, Min-Norm and Root-MUSIC, where $\hat{\mathbf{U}}_n$ is a $M \times (M-d')$ matrix in which the colume vectors are eigenvectors of $\hat{\mathbf{R}}$ corresponding to the $(M-d')$ minimum eigenvalues. $d' \leq d$ is the number of noncoherent signals. $\hat{\mathbf{P}}_s^\perp$ is the estimate of $\mathbf{P}_s^\perp$, the projective matrix onto the noise subspace. However, it is not difficult to prove that the projective matrix $\hat{\mathbf{P}}_s^\perp$ can equivalently be substituted for by the following $M \times M$ matrix $\hat{\mathbf{Q}}_T$ that only needs inversion computation (Marcos et al. 1990).

$$\hat{\mathbf{Q}}_T = \hat{\mathbf{Q}}(\hat{\mathbf{Q}}^H \hat{\mathbf{Q}})^{-1} \hat{\mathbf{Q}}^H$$

$$\hat{\mathbf{Q}} = \begin{bmatrix} \hat{\mathbf{P}} \\ \cdots \\ \mathbf{I} \end{bmatrix}_{M \times (M-d)}, \quad \hat{\mathbf{P}} = [\hat{\mathbf{F}}^H (\hat{\mathbf{G}} \hat{\mathbf{G}}^H)^{-1} \hat{\mathbf{G}}]^H$$

where $\mathbf{I}$ is a $(M-d) \times (M-d)$ identity matrix, $\hat{\mathbf{F}}$ and $\hat{\mathbf{G}}$ are partition matrices of $\hat{\mathbf{R}}$

$$\hat{\mathbf{R}} = [\hat{\mathbf{G}}_{M \times d} \vdots \hat{\mathbf{F}}_{M \times (M-d)}]$$

In a word, if we are able to calculate very fast a general equation as

$$\mathbf{V} = \mathbf{B} \mathbf{A}^+ = \mathbf{B}(\mathbf{A}^H \mathbf{A})^{-1} \mathbf{A}^H \quad 27$$

where $\mathbf{A}^+$ is the Moore-Penrose inversion, then all algorithms outlined above can be implemented in real time. For example, by letting $\mathbf{A} = \mathbf{B} = \mathbf{A}(\theta)$, we can implement WSF, DML and AM; by letting $\mathbf{A} = \hat{\mathbf{G}}^H$, $\mathbf{B} = \hat{\mathbf{F}}^H$ firstly, we get $\hat{\mathbf{P}}$ and $\hat{\mathbf{Q}}$, and then by letting $\mathbf{A} = \mathbf{B} = \hat{\mathbf{Q}}$, we can implement MUSIC, Min-Norm, and Root-MUSIC. With some small modifications we can also implement ESPRIT.

Some neural networks have recently been developed to calculate equation 27 very fast. The main idea is as follows. The matrix $\mathbf{V}$ in equation 27 can be regarded as the solution of a optimization problem

$$\min_{\mathbf{V}} \| \mathbf{B} - \mathbf{V}\mathbf{A} \|_F^2 \triangleq \min_{\mathbf{V}} Tr\{(\mathbf{B} - \mathbf{V}\mathbf{A})(\mathbf{B} - \mathbf{V}\mathbf{A})^H\} \quad 28$$

where $\|\cdot\|_F$ stands for the Frobenius norm. Suppose we have constructed a neural network with the output layer being a two-dimensional array of neurons so that the outputs can be represented by a matrix $\mathbf{V}(t)$. If we can express the Lyapunov energy function $E(t)$ of the whole network in a form approximate to $Tr\{(\mathbf{B} - \mathbf{V}(t)\mathbf{A})(\mathbf{B} - \mathbf{V}(t)\mathbf{A})^H\}$, where $\mathbf{A}$ and $\mathbf{B}$ are known matrices, and prove that $dE(t)/dt \leq 0$, i.e. the network is stable, we then have

$$E(t) \to E_{\min}$$
$$V(t) \to \underset{V}{\arg\min}\{Tr(B - VA)(B - VA)^H\} \quad \textit{approximately}$$

when $t \to \infty$. That means the final stable state of outputs of such a neural network is approximately the optimum solution we need. In a following example we will give a neural network which meets the requirements above and has a convergence time on the order of nanoseconds, which is short enough to guarantee the calculation of BA^+ in real-time. Before entering into details of such a neural network, however, we should note that the complex-valued matrix optimization problem in equation 28 could be changed to a corresponding real-valued one by simple algebraic operations. Moreover, let V, B, A be $p \times n$, $p \times m$ and $n \times m$ real-valued matrix respectively, and b_i be the i-th row vector of B, v_i be the i-th row vector of V. Equation 28 can be rewritten as:

$$\min_V \{Tr(B - VA)(B - VA)^T\} = \sum_{i=1}^{p} \min_{v_i}\{[b_i - v_i A][b_i - v_i A]^T\} \quad 29$$

which means that finding an optimum solution for matrix V is equal to finding optimum solution for vectors v_i separately and in parallel. The latter is a typical non-constraint Least Square optimization problem, and can be solved by several neural networks.

Illustrative Example — Parallel Tank-Hopfield Network Used to Implement Eigenvector-Based DOAs Estimation

Tank and Hopfield have proposed a neural network to solve the linear or quadric optimization problems of general form (Tank et al. 1986). When we consider the optimization problem

$$\min_{v_i}\{[b_i - v_i A][b_i - v_i A]^T\} \quad 30$$

a simplified version of Tank-Hopfield network is enough. The whole circuit structure is shown in Figure 10, which could be divided into the left part and the right part. The left part consists of n neurons with the output voltage denoted by $v_{ij}(t), j = 1, \cdots, n$. The right part consists of m neurons with the output voltage denoted by $\varphi_{il}(t), l = 1, \cdots, m$. The j-th neuron in the left part and the l-th neuron in the right part are shown in Figure 10(b). Wherein $-b_{il}$ is a constant current source, A_{jl} plays a role of cross-conductor, $g(\cdot)$ is a linear voltage amplifier, $f(\cdot)$ is a linear current-to-voltage device or a CCVS (current controlled voltage source). A concise representation for the whole structure is also shown in Figure 10(c). For each neuron j in the left part we have

$$C\frac{du_{ij}(t)}{dt} = -\frac{u_{ij}(t)}{R} - \sum_{l=1}^{m}\varphi_{il}(t)A_{jl}, \quad j=1,\cdots,n \qquad 31$$

$v_{ij}(t) = K_1 u_{ij}(t)$ and for each neuron l in the right part

$$\varphi_{il}(t) = K_2[\sum_{j=1}^{n}v_{ij}(t)A_{jl} - b_{il}], \quad l=1,\cdots,m \qquad 32$$

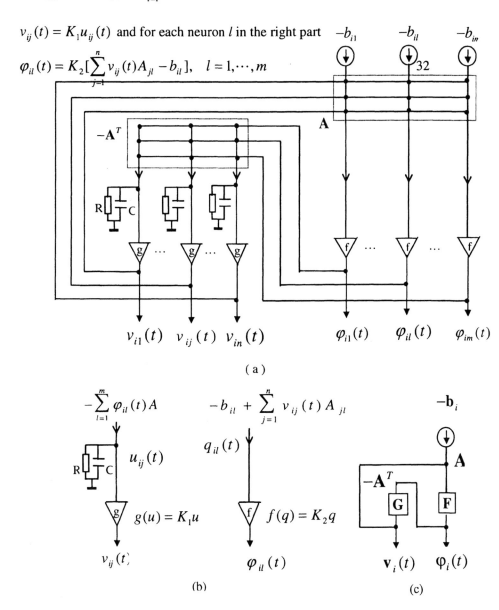

Figure 10 - The Simplified Version of Tank-Hopfield Network Used to Solve the LS Problem $\arg\min_{\mathbf{v}_i}\{[\mathbf{b}_i - \mathbf{v}_i\mathbf{A}][\mathbf{b}_i - \mathbf{v}_i\mathbf{A}]^T\}$. (a) the Whole Circuit Structure; (b) the j-th Neuron in the Left Part and the l-th Neuron in the Right Part; (c) the Concise Representation for the

The whole system described by equations 31 and 32 is a dynamic system. A problem we are concerned with is whether the system converges to a stable solution when time approaches to infinite. It can be answered by using the "energy principle". The basic concept of the energy principle may be expressed clearly but not very strictly as "*For any dynamic system if we are able to define a function E of the system state, which is bounded from below and always decrease with respect to any change of the system state, then the system must have a stable solution.*" Such a function E is called Lyapunov energy function. In order to prove that the simplified Tank-Hopfield network in Figure 10 does converge to a stable solution, we define a Lyapunov energy function as

$$E_i(v_{i1},\cdots,v_{in}) \triangleq \frac{K_2}{2}\sum_{l=1}^{m}\{b_{il}-\sum_{j=1}^{n}v_{ij}(t)A_{jl}\}^2 + \sum_{j=1}^{n}\frac{1}{2RK_1}v_{ij}^2(t)$$

It is obvious that $E_i \geq 0$ and, by making use of equation 31 and 32,

$$\frac{dE_i}{dt} = -\sum_{j=1}^{n}\frac{C}{K_1}\left[\frac{dv_{ij}(t)}{dt}\right]^2 \leq 0$$

which means E_i is bounded from below and always decrease with respect to any change of v_{ij} and $dE_i/dt = 0$ when and only when $dv_{ij}(t)/dt = 0$. According to the energy principle, the time evolution of the network state $v_{ij}(t)$ will attempt to seek out the minimum of E_i and comes to stop at such points in state space, which corresponds to the condition $dv_{ij}(t)/dt = 0$. Thus the stable solution can be obtained from equation 31 by letting $dv_{ij}(t)/dt = 0$, i.e.

$$\lim_{t\to\infty}v_{ij}(t) = v_{ij} = -RK_1\sum_{l=1}^{m}\varphi_{il}A_{jl}$$

$\varphi(t)$

or

$$\mathbf{v}_i = -RK_1 \varphi_i \mathbf{A}^T \qquad 33$$

Where φ_i is a row vector with element φ_{il}. The vector form of equation 32 is

$$\varphi_i = K_2[\mathbf{v}_i \mathbf{A} - \mathbf{b}_i] \qquad 34$$

Combining equation 33 and 34 results in

$$\mathbf{v}_i = -RK_1 K_2 \mathbf{v}_i \mathbf{A}\mathbf{A}^T + RK_1 K_2 \mathbf{b}_i \mathbf{A}^T$$

$$\Rightarrow \mathbf{v}_i [\mathbf{A}\mathbf{A}^T + \frac{1}{RK_1 K_2} \mathbf{I}] = \mathbf{b}_i \mathbf{A}^T$$

$$\Rightarrow \mathbf{v}_i = \mathbf{b}_i \mathbf{A}^T [\mathbf{A}\mathbf{A}^T + \frac{1}{RK_1 K_2} \mathbf{I}]^{-1}$$

$$\Rightarrow \mathbf{v}_i \approx \mathbf{b}_i \mathbf{A}^T [\mathbf{A}\mathbf{A}^T]^{-1} = \mathbf{b}_i (\mathbf{A}^T \mathbf{A})^{-1} \mathbf{A}^T = \mathbf{b}_i \mathbf{A}^+ \qquad 35$$

which is just the pseudo-inversion solution of the optimization problem in equation 30. The approximation error in equation 35 can be made arbitrarily small by using large enough K_1, K_2 and R. We now return to the more general optimization problem in equation 29. As noted before, each row vector $\mathbf{v}_i$ can be optimized separately, so we can combine p simplified Tank-Hopfield networks in parallel to achieve the optimization of $\mathbf{V}$. The corresponding circuit structure and its concise representation are shown in Figure 11. To start up the network, we use $\mathbf{A}$ as $n \times m$ weight matrix from main-network outputs to sub-network inputs, $-\mathbf{A}^T$ as $m \times n$ weight matrix from sub-network outputs to main-network inputs, and apply $-\mathbf{B}$ as $p \times m$ input current source matrix of the sub-network. Hereafter the system operates in a very complicated way and finally approaches to a unique stable state,

$$\mathbf{V}(t) \rightarrow \mathbf{V} = \begin{bmatrix} \mathbf{v}_1 \\ \vdots \\ \mathbf{v}_p \end{bmatrix} \approx \begin{bmatrix} \mathbf{b}_1 \mathbf{A}^+ \\ \vdots \\ \mathbf{b}_p \mathbf{A}^+ \end{bmatrix} = \mathbf{B}\mathbf{A}^+$$

The convergence time can be quantitatively found out from the eigenvalues of $\frac{1}{RC}\mathbf{I}+\frac{K_1 K_2}{C}\mathbf{AA}^T$. Because $\mathbf{AA}^T$ is non-negative, the smallest eigenvalue $\geq \frac{1}{RC}$. For example, when using the typical parameters $K_1=10, K_2=10, R=1k\Omega, C=100pF$, the largest characteristic time constant $\leq RC = 100\ ns$. In conclusion, the proposed neural network can provide the fast calculation ability necessary for the real time implementation of the eigenvector-based DOAs estimation within an elapsed time of only a few characteristic time constants (on the order of hundred nanoseconds).

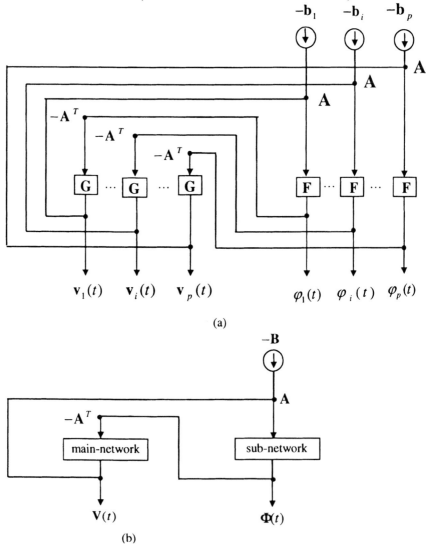

Figure 11 - The Parallel Tank-Hopfield Networks Used to Implement the LS Problem $\arg\min_{\mathbf{V}} Tr\{(\mathbf{B}-\mathbf{VA})(\mathbf{B}-\mathbf{VA})^T\}$. (a) The complete Circuit Structure ;
(b) The Concise representation of the system

5. REFERENCES

1. Broomhead, D. S., Jones, R., McWhirter, J. G. and Shepherd, T. J. (1989). " A Systolic Array for nonlinear Multi-Dimensional Interpolation Using Radial Basis Functions", *Electronic Letters*, 26, 7-9.

2. Di Zitti, E., Bisio, G. M., Caviglia, D. D. and Parodi, G. (1989). "Neural Networks on a Transputer Array", in *Proc. ICASSP'89*, 2513-2516.

3. Di Zitti, E., Caviglia, D. D., Bisio, G.M. and Parodi, G. (1989). "Analysis of Neural Algorithms for Parallel Architectures", *Proc. Int. Symp. on Circuit and Systems*, Protland.

4. Golub, G. H. and Vanloan, C. F. (1980). "An Analysis of the Total Least Squares Problem", *SIAM J. Num. Anal.*, 17, 883-893.

5. Haykin, S. (1994). *Neural Networks: A Comprehensive Foundation*, New York: Macmillan College Publishing Company.

6. Haykin, S. (1996). "Neural Networks Expand SP's Horizons", *IEEE SP Magazine*, 13, 24-49.

7. Hopfield, J. J. and Tank, D. (1985). "Neural computation of Decisions in Optimization Problems", *Biol. Cybern.*, 52, 141-152.

8. Hush, D. R. and Horne, B.G. (1993). "Progress in Supervised Neural Networks: What's New since Lippmann?", *IEEE SP Magazine*, 10, 8-39.

9. Krim, H. and Viberg, M. (1996). " Two Decades of Array Signal Processing Research", *IEEE SP Magazine*, 13, 67-94.

10. Kung, S. Y. (1985). "VLSI Array Processor for Signal Processing ", in *Modern Signal Processing* edited by T. Kailath, Springer-Verlag, 393-440.

11. Lippmann, R. P. (1987). "An Introduction to Computing with Neural Nets", *IEEE ASSP Magazine*, 4, 4-22.

12. Luo Falong , Bao Zheng and Zhao Xiaopeng (1992). "Real-Time Implementation of 'Propagator' Bearing Estimation Algorithm by Use of a Neural Network", *IEEE J. Oceanic Engineering*, 17, 4, 320-325.

13. Marcos, S. and Benidir, M. (1990). "On a high Resolution array Processing Method Non-Based on the Eigenanalysis Approach", in *Proc. ICASSP'90*, 2955-2958.

14. Ottersten, B., Viberg, M. and Kailath, T. (1992). "Analysis of Subspace Fitting and ML Techniques for Parameter Estimation from Sensor Array Data", *IEEE Trans. SP*, 40, No.3, 590-600.

15. Owsley, N. L. (1985). "Sonar Array Processing ", in *Array Signal Processing* edited by S.Haykin, Prentice Hall, Englewood Cliffs, New Jersey, 115-193.

16. Schmidt, R. O. (1986). "Multiple Emitter Location and Signal Parameter Estimation", *IEEE Trans. Antennas Propaga.*, 34, 276-280.

17. Tank, D. and Hopfield, J. J. (1986). "Simple 'Neural' Optimization Networks: An A/D Converter, Signal Decision Circuit, and a Linear Programming Circuit", *IEEE Trans. CAS*, 33, No.5, 533-541.

18. Viberg, M. and Ottersten, B. (1991). "Sensor Array Processing Based on Subspace Fitting ", *IEEE Trans. SP*, 39, No.5, 1110-1121.

19. Zhang Chun-yuan, Dai Kui and Hu Shou-ren (1995). "Neuro C: A General Purpose Parallel Neuro-Computer", in *Proceeding of IEEE Intenational Conference on Neural Networks and Signal Processing* Vol. I, Southeast University, Nanjing, China, 156-158.

20. Ziskind, I. and Wax, M. (1988). "Maximum Likelihood Localization of Multiple Sources by Alternating Projection", *IEEE Trans. ASSP*, 36, No.10, 1553-1560.

6

Statistical Signal Processing of Echo Ensembles

J.D. Penrose* and T. Pauly**
Centre for Marine Science and Technology, Curtin University of Technology, Kent St., Bentley 6102, Western Australia

Australian Antarctic Division, Channel Hwy., Kingston 7050, Tasmania**

1. INTRODUCTION

This chapter is concerned with gaining information about acoustic targets from ensembles of backscattered acoustic signals. Emphasis is given to backscatter from point targets, such that individual acoustic cross sections are small compared with the cross sectional area of the interrogating sound beam at ranges of interest. Biological scatterers in the ocean usually constitute such point targets and are used here to provide an example of ensemble processing. Such processing calls for the adoption of a model describing aspects of the scattering and the development of processing techniques. The assumptions involved in the model and processing technique used are of central importance. In this description, attention is given to the use of Monte Carlo techniques to evaluate the assumptions employed and the robustness of the techniques described.

The analysis that has been undertaken in relation to the use of acoustics in marine biomass estimation is very substantial, with an extensive literature covering theoretical and experimental work and the relationships between them. Acoustic devices ranging from simple echo sounders to sophisticated echo counting and integration systems are in use for purposes of biological resource estimation. One important problem in such acoustic work concerns the accurate estimation of target strength for in-situ targets. Such estimation is of importance because of the role target strength values play in assessing total biomass and also because target strength can be used in some applications as a first, albeit crude, method of distinguishing target types. In this description, the major aim of the processing is to derive estimates of the acoustic target strength of an individual target from a large number of measured backscattered signals. Each individual measured signal arises ideally from only one target, but is subject to two separate processes which modify the form and magnitude of the signal. These processes, the angular position of the target in the beam and the attitude or orientation of the target to the interrogating sound pulse are independently random in nature. The signal processing methodology described in this chapter incorporates this randomness in a suite of assumptions which are used in forming the processing technique.

2. MONOSTATIC SOUNDING OF SINGLE POINT TARGETS

Figure 1 represents scattering from a single target at range r from a simple disk transducer, sounding into its far field only. This monostatic geometry is representative of the configuration most commonly available in marine sounder installations. It provides less information than that available from split and dual beam installations but offers advantages in terms of cost and simplicity. Also, a single transmitting / receiving transducer will be smaller that an equivalent split or dual beam unit and thus, for a given frequency, will have a smaller near-far field transition distance, an issue of significance for very small scatterers. Backscattered signals are most readily interpreted for any transducer configuration, when they come from interactions in the far field rather than the near field region.

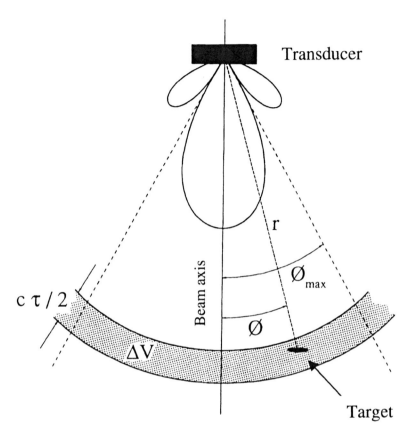

Figure 1 - Monostatic Sounding Geometry

In the scattering geometry shown in figure 1 the transducer has emitted a pulse of length τ in the time domain and consequently $c\tau$ in the space domain. At some range r a single target is located at angle ϕ to the beam axis. The single target specification is important in the analysis that follows. Note that the idealised transducer has a beam pattern symmetric about the beam axis with sensitivity $b(\phi)$ in both transmission and reception given by;

$$b(\phi) = \frac{p(\phi)}{p(0)} \qquad \qquad 1$$

The "range slice" associated with a received pressure level at a specific time t after pulse emission at t = 0 is centred at a distance ct/2 away from the transducer and has thickness ct/2. The setting of pulse length τ also sets the limit of the resolution of target separation that the system can provide. Here c is the sound speed. It is possible to enlarge the slice thickness by accumulating successive cτ/2 thicknesses within gated ranges, say r_1, to r_2.

Consider r sufficiently large that the Fraunhofer region is involved. The received pressure back at the transducer, p_s may be expressed as;

$$p_s = \frac{r_0 p_0}{r} e^{-\alpha r} b(\phi)(\sigma_{bs})^{0.5} \frac{e^{-\alpha r}}{r} b(\phi)$$

transmission   reception

where α is the absorption coefficient.

Hence;

$$p_s = \frac{r_0 p_0}{r^2} e^{-2\alpha r} b^2(\phi)(\sigma_{bs})^{0.5} \qquad 2$$

where p_0 is the source pressure measured on axis at distance r_0 from the transducer. The acoustic cross section of the target is given by;

$$\sigma_{bs} = \left(\frac{p_{bs}}{p_p}\right)^2 r^2 \qquad 3$$

where, p_{bs} is the sound pressure measured at distance r away from a target on which sound pressure p_p is incident, it will be useful to put r = 1 in equation 3 so that;

$$(\sigma_{bs})^{0.5} = \frac{p_{bs}}{p_p} \qquad 4$$

Under these conditions;

$$p_p = \frac{r_0 p_0}{r} e^{-\alpha r} b(\phi) \qquad\qquad 5$$

$$p_{bs(r=1m)} = p_p (\sigma_{bs})^{0.5} \qquad\qquad 6$$

Two features of equation 2 have random characteristics in many scattering applications. If targets as shown in figure 1 are randomly distributed in space, the distribution of $b(\phi)$ values may be specified on an ensemble basis. The volume of the range slice illustrated in figure 1 may be approximated by;

$$V \approx \pi (r\phi_{max})^2 \frac{c\tau}{2} \qquad\qquad 7$$

If the average global density of targets is ε per unit volume then the number of targets insonified in V is εV. This number, insonified at range r per pulse transmission is dependent on the probability that, for a given average target density (expressed over a larger system volume very much bigger than V) one or more targets will fall in the designated volume V. The Poisson distribution gives the probability P that the number of targets N in V is k, where

$$P_{(N=k)} = \frac{(\varepsilon V)^k e^{-\varepsilon V}}{k!} \qquad\qquad 8$$

On average, εV targets are expected to appear in volume V and this figure can be used to decide when the appearance of a multiple targets in the same minimum range slice of thickness $c\tau/2$ can be expected to take place with sufficient frequency to negate signal processing techniques which rely, as in this case, on scattering from one target at a time. In general if $<\varepsilon V>$ is <0.2, the predominance of single scatterer occupancy of V is essentially complete (Palumbo et al. 1993). This work has shown in a Monte Carlo simulation of such processes that for some purposes $<\varepsilon V>$ as high as 0.6 may be tolerated under such conditions. This means that P(0), the probability of there being no target in the range slice is dominant, P(1) is relatively high, P(2) is low and P(3) is very low indeed. Since the signal processing can be structured to ignore "no signal returns" (i.e. those described by P(0)), and since P(1) is the largest of the non-zero probabilities, the influence of P(2) and P(3) can be ignored.

When single scatterer events predominate, the mode of signal processing described in this chapter is appropriate. When multiple scatterer occupancy begins to dominate, as ε increases (or, for a given ε, as V increases), it is necessary to switch to echo integration in which,in the simplest case, the scattered energy from a number of scattering centres of equal cross section is proportional to the number of scatterers insonified. In marine applications, the total energy scattered is linked to the biomass insonified, a consequence of the incoherent nature of such scattering.

At still higher global scatterer densities, issues of multiple scattering and shadowing begin to become important, As noted above, in the present treatment, we consider only the global scatterer density range which allows single scatter interactions to dominate.

Consider firstly the development of the $b(\phi)$ term appropriate to randomly positioned scatterers. Under these conditions, the number of targets in the angular field ϕ to $\phi+d\phi$ increases with ϕ. The probable number of targets in the volume element ΔV is;

$$N = \varepsilon \Delta V$$

$$= \varepsilon 2\pi r \phi_i r d\phi_i \frac{c\tau}{2} \qquad\qquad 9$$

$$= \varepsilon \pi \phi_i c \tau r^2 d\phi_i$$

Since ΔV increases linearly with ϕ_i, the probable number of targets should increase linearly with ϕ. Thus, for random target distributions, the N values which will occur due to this distribution should increase up to the maximum that the beam geometry will allow.

The second component of equation 2 which may be stochastic in form is the target descriptor $\sigma_{bs}^{0.5}$. The relationship between insonifying wavelength λ, often expressed as the related wavenumber $k = 2\pi/\lambda$, and some parameter describing target size, becomes important. The key issues are illustrated by reviewing some of the results arising from the classic treatment by Rayleigh of acoustic scattering from non-resonant compliant spheres. When $\lambda \gg a$, where a is the sphere radius, Rayleigh showed that the sound scattering varies as the fourth power of the wave number, i.e. as the inverse fourth power of the wavelength. Backscatter from such target spheres, while dependent on the contrast in density and elasticity between target and scattering medium, would not be modified by changes in the target attitude. This result is expected to remain valid even if the target is not spherical, so

Chapter 6 211

long as λ remains much larger than the principal dimension of the scatterer, i.e. while the conditions of Rayleigh scattering are sustained. Thus, for a target population of idealised spheres in the Rayleigh region, σ_{bs} can be modelled as a constant.

In the Rayleigh scattering regime, values of σ_{bs} are necessarily low. This means that in oceanographic sounding practice, Rayleigh scatterers will often not provide sufficient backscatter to allow signals from individual targets to be resolved. Larger targets, providing larger σ_{bs}, can be individually detected at larger r values. As the value of ka rises, in the case of a sphere, to around ka = 1, backscatter behaviour enters the geometrical region, where σ_{bs} becomes more constant with ka, but now depends in detail on target attitude in the sound beam. Under these conditions, the detailed value of σ_{bs} may be considered to arise from the superposition of backscattered signal components from a number of sites within and distributed along the target. The simplest model of such behaviour considers the target to be comprised of a number of scattering centres which contribute randomly phased components to the total backscatter signal. When a large number of signal components of equal magnitude but random phase are summed a large number of times, the resultant ensemble of values will be represented by the Rayleigh distribution. For convenience, we adopt the concept of an instantaneous "scattering length" l, where;

$$\langle l^2 \rangle = \sigma_{bs} = (l_\sigma)^2 \qquad \qquad 10$$

Here the angle brackets indicate an ensemble average. The parameter l_σ is thus equal to the square root of the backscattering cross section.

Expressed in scattering length terms, the Rayleigh distribution becomes;

$$w_R(l) = \frac{|l|}{\langle l^2 \rangle} e^{\frac{-l^2}{2\langle l^2 \rangle}} \qquad \qquad 11$$

Thus, as a first approximation to the scattering from geometric region scattering from individual targets, equation 2 involves two terms with stochastic characteristics. For targets randomly distributed in space, the values of ϕ_i expected arise randomly, but on average, the probability of occurrence will increase linearly with ϕ up to a limit set by the beam geometry. The values of scattering length l will also arise randomly. An ensemble of scattering lengths will, as a first approximation, be expected to follow a Rayleigh distribution. For a calibrated sounder system, the remaining terms in equation 2 are either known or measurable from the time-domain record. It is thus in principle possible to simulate an ensemble of p_s values, as returned by equation 2, using Monte Carlo techniques. Such simulation has value in treating certain classes of field data, and for testing signal processing techniques.

3. MONTE CARLO SIMULATION

Values of the two stochastic parameters ϕ and l need to be simulated so that a large number of realisations follow the linear and Rayleigh probability of occurrence trends outlined above.

Let u be a variable with probability density function f(u). We define another variable v = g(u) and seek the appropriate probability density function h(v), such that;

$$h(v) = f(u)\frac{du}{dv} \qquad 12$$

This may be considered to represent the mapping from the u to v axis, of an input distribution f(u) to produce a new distribution h(v), by reflection from the gradient $\frac{du}{dv}$ of the function v = g(u). The requirement is therefore, for each distribution sought, to establish the appropriate form of v = g(u). It is also necessary to adopt an input distribution f(u). A suitable candidate distribution is one where the probability of occurrence of u is uniform over some selected interval, e.g. specify u = (RND) to provide a distribution uniform over the interval 0 - 1.

Consider the requirement that an ensemble of v values be generated such that the probability that a simulated value of v will occur increases linearly with v. This requires that;

$$\frac{du}{dv} = C_1 v \qquad 13$$

where C_1 is a constant. Thus,

$$du = C_1 v dv$$

and

$$u = \frac{C_1 v^2}{2}$$

Put $C_1 = 2$, hence:

$$v = u^{0.5} \quad \text{or} \quad v = (RND)^{0.5} \qquad 14$$

To represent the figure 1 geometry, and putting $v = \phi_i$ modifies equation 14 to become;

$$\phi_i = \phi_{max}(RND)^{0.5} \qquad 15$$

Now consider the requirement that an ensemble of v values be generated to follow a Rayleigh distribution, i.e. v values distributed according to

$$\frac{2v}{n} e^{-\frac{v^2}{n}},$$

thus;

$$\frac{du}{dv} = \frac{2v}{n} e^{-\frac{v^2}{n}} \qquad 16$$

where $n = 2m^2$, m being the mode or most common value of the distribution. It can be shown (Palumbo et al. 1993) that;

$$\sigma_{bs} = (l_\sigma)^2 = \langle l^2 \rangle = n \qquad 17$$

From 16;

$$u = \frac{2}{n} \int v e^{-\frac{v^2}{n}} dv$$

and

$$u = 1 - e^{-\frac{v^2}{n}}$$

so that

$$v = (-n \ln(1-u))^{0.5}$$

To suit the simulation of scattering lengths, this becomes;

$$l = (-n \ln(1 - RND))^{0.5} \quad \text{or} \quad l = (-n \ln(RND))^{0.5} \qquad 18$$

From ϕ_i values generated using equation 15 the related values for $b(\phi)$ may be produced. For a simple disk transducer of radius a, the value for $b(\phi)$, the pressure directivity is given by;

$$b(\phi_i) = \frac{2 J_1(ka \sin \phi_i)}{ka \sin \phi_i} \qquad 19$$

Equation 17 can be evaluated by using a polynomial approximation e.g. Abramowitz and Stegun (1965).

Equations 2, 16, 18 and 19 indicate that for the conditions specified, it is possible to simulate by Monte Carlo methods an ensemble of backscattered signals from single targets. In the formalism presented, only one fitted parameter, n, is called for. This parameter, introduced in equation 16 is, as shown in equation 17, a surrogate for σ_{bs}. The possibility thus exists to extract estimates of σ_{bs} from an ensemble of echo returns from the sounder geometry shown in figure 1. In this regard, it is useful to define the term Target Strength (TS). Considering mean values yields;

$$\overline{TS} = 10 \log \overline{\sigma_{bs}} \qquad 20$$

4. TARGET STRENGTH ESTIMATION FROM ECHO ENSEMBLES

Sections 9.1 and 9.2 show how for an idealised sounder configuration, ensembles of backscattered signals may be simulated. Further, the simulation process points to the possibility that a suitable signal processing process might enable such simulated ensembles to be compared with field data and thus yield TS estimates for the target population. This method of TS estimation has been termed by Foote (1991) the "Indirect" approach to Target Strength determination. Its origins can be traced to Craig and Forbes (1969) and the technique has been applied by Peterson *et al.* (1976), Ehrenberg *et al.* (1981) and, as described below, by Pauly and Penrose (1998). The summary below draws on the work of Peterson *et al.* and Palumbo *et al.* (1993).

The application of the Indirect technique to TS estimation necessarily involves a series of assumptions. Consider the following set of assumptions;

a) A single species of target is present and all targets have the same σ_{bs}
b) The targets are located randomly in space
c) Only one target is insonified at a time in the range cell.
d) The targets are large enough to be in the geometric scattering region for the sound wavelength used. Fluctuations in backscattered pressure amplitude due to changes in target attitude in the beam are represented by a Rayleigh distribution.
e) The transducer beam pattern is accurately known.

These assumptions, invoked in sections 9.1 and 9.2 above, were used by Peterson *et al.* to match theoretically derived probability density functions with frequency of occurrence data from a fisheries survey undertaken in Lake Michigan.

A first step in data analysis is to define a normalised echo amplitude e for each backscattered return gained in the field measurements. In the terms used in equation 2;

$$e = p_s r^2 10^{\frac{\alpha r}{10}} / p_0 r_0 \qquad 21$$

so that;

$$e = lb(\phi) \qquad 22$$

Thus e is the product of the two statistically varying parameters, l and $b(\phi)$, discussed above. Fluctuations in these two parameters are described by the probability density functions $w_R(l)$ and $w_I(b)$ so that the combined probability function $w_E(e)$ may be defined as;

$$w_E(e) = \int_0^1 \frac{w_I(b) w_R(e/b)}{b} db \qquad 23$$

Plotting this PDF as $w_E(e)(ka^2)l_\sigma$ vs e/l_σ renders both axes dimensionless. Figure 2(a) illustrates the process by which the normalised data distribution is compared with the equivalent theoretical distribution. This fitting technique, termed by Palumbo et al. the "PDF Method" calls for the data to be binned in order to develop the frequency-of-occurrence histogram shown. This process reduces the quality of the data-theory fit obtained. Palumbo et al. developed an improved technique, termed the "Ranked Array Method" which is significantly more accurate and faster to implement. This calls for use of the integral form of equation 23, the cumulative distribution function of the normalised echo amplitude e;

$$C(e) = \int_0^1 w_I(b) \left(\int_0^{e/b} w_R(l) dl \right) db \qquad 24$$

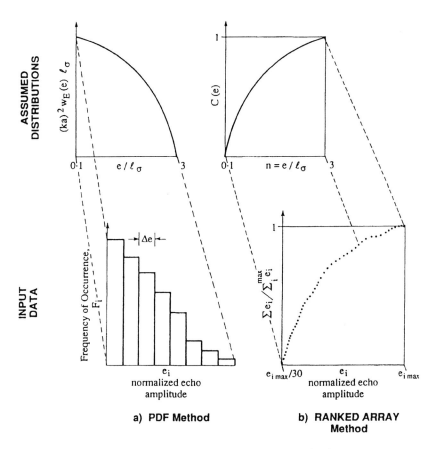

Figure 2 - PDF and Ranked Array Methods

As shown in figure 2(b) this is then compared with the equivalent experimental values; i.e.,

$$\frac{\sum e_i}{\sum_1^{max} e_i} \qquad 25$$

In this technique all incoming experimental data are ranked in amplitude order from the smallest value (e_1) to the largest (e_{max}) and summed from e_1 to e_i, so that when e_{max} is reached the normalised cumulative distribution has value unity. Both PDF and Ranked Array Methods need to specify a lower threshold to exclude noise dominated signals. In the figure 2(b) example this has been set at $e_{max}/30$, corresponding to a dynamic range, smallest-to-largest signal received, of 30. In the Ranked Array Method, fitting between the theoretical and experimental curves calls only for adjustment of the abscissa and curve matching is undertaken at all points on the cumulative distribution (Pauly and Penrose, 1998).

Both the PDF and Ranked Array Methods as presented here depend on the applicability of the assumptions a – f listed above. Palumbo *et al.* (1993) have carried out an extensive set of tests to evaluate the influence on derived TS values of departures from the idealised conditions associated with the assumptions. The procedure involved generating simulated data prepared under various departures from ideality and comparing derived TS values determined from this data using the Ranked Array Method with the original values used in the simulation process. A key assumption relevant to most ocean conditions is assumption a. Even in the comparatively low bio-diversity environment of Antarctica, where large assemblies of essentially mono-specific organisms occur, significant variation in target size and TS is to be expected within a sampled population. Palumbo *et al.* addressed this issue by generating data using TS values appropriate for a population with a Gaussian distribution of lengths. In a second suite of tests, populations comprising significantly differing sizes were used in data simulation. Each of the other assumptions noted above was evaluated in a related fashion. Table 1 summarises the results of this evaluation process.

Assumption d, concerning the applicability of the Rayleigh distribution, has received considerable attention. Much of the discussion is summarised in Medwin and Clay (1998).

5. A CASE STUDY – THE TARGET STRENGTH OF ANTARCTIC KRILL

Pauly and Penrose (1998) have used the Ranked Array Method in the evaluation of the Target Strength of Antarctic krill, *Euphausia superba*, in a laboratory test tank. Figure 3 shows the ~10 m^3 stainless steel tank which was filled with sea water and maintained within a temperature range of – 0.5 to 1.0 C.

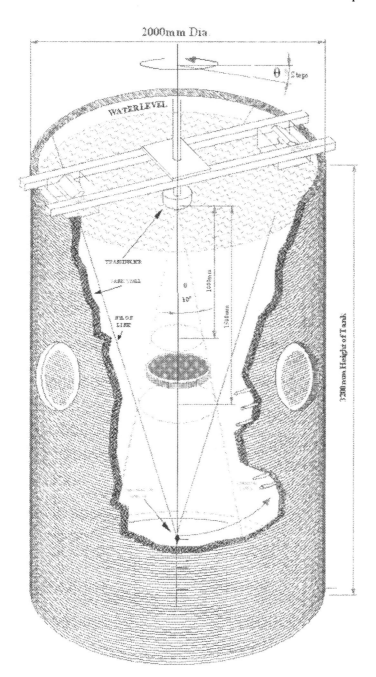

Figure 3 - Experimental Tank Configuration

A 120 kHz monostatic sounding system was developed, using the transducer shown near the top of the tank in figure 3. The received signal was gated to provide a sampling volume starting at a range of $r_1 = \sim 1$m and extending to $r_2 = \sim 1.5$m from the active face of the transducer. A number of krill, ranging from 22 to 48 over 9 experiments, were introduced into the tank to provide a global target density ε such that the probability of P(1) as described by equation 8 greatly exceeded P(2) and P(3). A consequence of this requirement, as discussed in section 9.1 is that P(0) is dominant, so that most signals emitted from the transducer did not encounter a target in the sampled volume. The experimental system was set to trigger when any target was encountered, and later post editing was carried out to remove the comparatively small proportion of P(2) and higher events. The post processing used measures of pulse shape and length. This technique, and the monitoring of phase consistency within the pulse, offer effective single target discrimination capability. Thus, assumption c noted in section 9.3 above was realised to good approximation. The controlled environment of the tank enabled targets to be selected to represent specific size classes, so that assumption a noted above in section 9.3 was closely approximated. Figure 3 also shows a stereo camera pair which was also triggered on detection of a target in the sample volume. Careful beam calibration was carried out, justifying the adoption of assumption e.

Processing of received backscatter signals proceeded as represented in figure 2(b) and associated discussion. A fuller description of the processing is found in Pauly and Penrose (1998). Since, however, Antarctic krill present small acoustic cross sections, the dynamic range of e values shown in figure 2(b), i.e. 30, was not realisable. The useable dynamic range defined by e_{max}/e_1, where e_1 is, as before, the smallest useable signal, varied in the range 2.59 – 5.19 for the krill targets. This issue introduces the concept of "threshold bias" discussed below. The experimental configuration allowed for substantial numbers of scattered signals to be accumulated. The numbers of signal events accumulated, i.e. the ensemble size, varied from 111 to 2254 over the 9 experiments conducted.

Figure 4 shows the results from experiment 4, which used an ensemble of 2254 scattered signals. The figure shows the experimental and theoretical cumulative distributions and in addition the TS values computed at each experimental data point. The variation in TS values seen, of approximately 1 dB thus corresponds to uncertainties introduced in measurement, the assumptions adopted and the use of the Ranked Array processing technique in this application. The final TS values produced from the data ensemble took into account the influence of the dynamic range of the available signals. Variations in this range can be expected to modify the performance of the Ranked Array or similar ensemble processing techniques. As the dynamic range decreases, the time required to collect a given number of samples increases. Thus, the likelihood of populating the larger e valued region of the distribution increases. That is, a larger sample is required to obtain a given number of e values above the threshold, increasing the likelihood of obtaining larger values and hence biasing the TS estimate. This effect was identified by Weimer and Ehrenberg (1975) as threshold bias. A Monte Carlo technique was used to evaluate this effect. Simulated data ensembles were generated using a known input Target

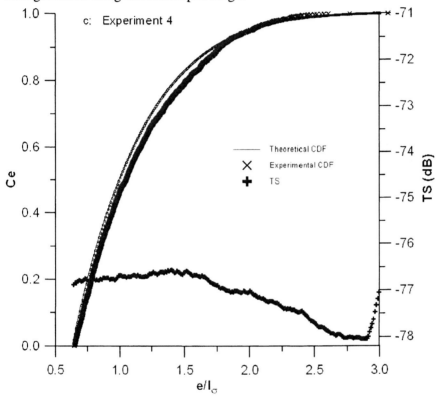

Figure 4 - Comparison of theoretical and experimental distributions

Strength (TS_{in}). Figure 5 shows the results of Ranked Array processing of the simulated data set, for a dynamic range in e ranging from 3 – 12.5. The performance of the processing is illustrated by evaluating the difference between the output Target Strength estimate TS_{out}, and TS_{in}. The threshold bias effect is seen to provide up to 1 dB over estimate at the lowest dynamic range shown, with decreasing influence as the dynamic range increases. Figure 5 also shows the form of a parameterised curve used to fit the trend shown, and to correct Ranked Array results for the threshold bias effect. Figure 6 shows the final TS values from the 9 experiments, as a function of target length, and compared with recent results from other workers using different experimental techniques.. The full line represents the form of the TS vs length relationship now adopted by the Scientific Committee of the Commission for the Conservation of Antarctic Marine Living Resources (CCAMLR) for Antarctic krill. The dotted line illustrates the relationship which was previously employed. The transition between the two relationships corresponds to a very significant increase in the estimation of krill populations in Antarctic waters.

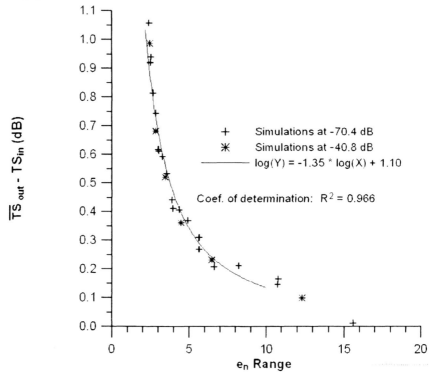

Figure 5 - Threshold bias vs signal dynamic range

The determination of the acoustic Target Strength of Antarctic krill illustrates the use of ensemble statistics to derive information on single scatterers. In this example, the assumptions invoked to establish a theoretical distribution are largely justified. The derived Target Strength values differ by only approximately 1 dB over the dynamic range of the distribution fitting available. This figure is comparable to uncertainties arising from calibration of field equipment. Further refinement is likely to call for attention to several of the assumptions invoked, notably the adoption of the Rayleigh distribution to account for signal variations due to target attitudes in the sound beam.

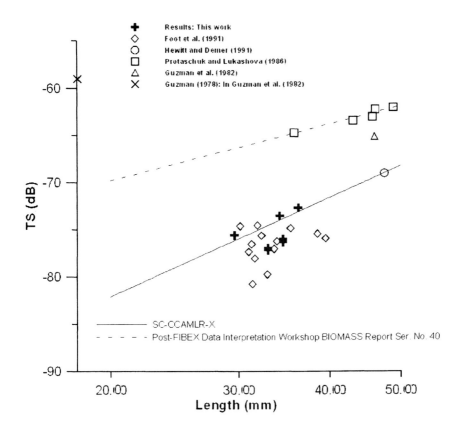

Figure 6 - Krill target strength vs length

6. REFERENCES

1. Abramowitz, M and Stegun, I. E. (eds) (1965). *Handbook of mathematical functions,* Dover, New York.

2. Craig, R.E. and Forbes, S.T. (1969). "Design of a sonar for fish counting", FiskDir. Skr. Ser. Havaunders, 15, 210-219.

3. Ehrenberg, J.E., Carlson, J.J., Traynor, J.J. and Williamson, N.J. (1981). "Indirect measurement of the mean acoustic scattering cross section of fish", J. Acoust. Soc. Am. 69, 955-962.

4. Foote, K.G. (1991). "Summary of methods for determining fish target stength at ultrasonic frequencies", ICES J. Mar. Sci. 48, 211-217.

5. Medwin, H. and Clay, C. S. (1998). *Fundamentals of acoustical oceanography,* Academic Press, London

6. Palumbo, D., Penrose, J.D. and White, B. A. (1993). "Target strength estimation from echo ensembles", J. Acoust. Soc. Am. 94, 2766-2775.

7. Pauly, T. and Penrose, J.D. (1998). "Laboratory target strength measurements of free swimming Antarctic krill (*Euphausia superba*)", J. Acoust. Soc. Am. 103, 3268-3280.

8. Peterson, M.L., Clay, C.S. and Brandt, S.B. (1976). "Acoustic estimates of fish density and scattering function", J. Acoust. Soc. Am., 60, 618-622.

9. Weimer. R.T. and Ehrenberg, J.E. (1975). "Analysis of threshold-induced bias inherent in acoustic scattering cross section estimates of individual fish", J. Fish. Res. Bd. Can., 32, 2547-2551.

ASSUMPTION	MODIFICATION OF ASSUMPTION	CONCLUSION
All targets have the same average dorsal cross section. (i.e. the same length)	Lengths distributed normally.	Target strength estimates remain within 1 dB of ideal, for coefficients of variation up to 20-30%
Only one size class of target is present.	Bimodal length – Frequency Distribution of targets. $L_{small} = 0.1056$ m $L_{large} = 0.264$ m	For population ratio of large/small targets ≥ 15, TS. estimate corresponds to larger targets. For population ratio large/small $\geq 1/200$ $\rightarrow 1/100$ TS. estimate corresponds to smaller targets.
Single target in the range shell at any one time.	Poisson distributed probability of several targets in range shell. $$P = \frac{(\varepsilon V)^T e^{-(\varepsilon V)}}{T!}$$ P = probability that no. of targets in volume V is T. A = average no. of targets/unit volume.	TS, within 1 dB for $\varepsilon V = 0.6$ i.e. for $\dfrac{P(2)}{P(1)} = 30\%$ and $\dfrac{P(3)}{P(1)} = 6\%$
Target orientation factor described by Rayleigh Distribution.	Distorted frequency distribution generated.	TS. Within 1 dB for significant distortions of the Rayleigh Distribution.
Transducer Beam Pattern is known and well defined.	Various beam pattern deformations undertaken.	Mild variations in beam pattern acceptable.

Table 1 - Simulation Summary

7

Advanced coding for Underwater Communication

Hui Junying, L Liu, Feng Haihong and Liu Hong
(Underwater Acoustics Institute, Harbin Engineering University, Harbin 150001)

1. INTRODUCTION

IT is well known that the recent advances in satellite communication and mobile communication technologies are not paralleled with same developments for digital underwater acoustic communication systems. In particular recent literature on underwater acoustic communications [1]-[21] deal extensively with the challenges in high speed underwater acoustic communications (see also chapter one of this book). However, although high speed underwater acoustic communication techniques for vertical channels in deep water environments are well investigated, there is no extensive studies on the horizontal channel case and the limitation of the multipath effects in this case. This is because in the horizontal channel case, the communication rate is low and error bit rate is high which add further complexity and challenge to apply the same vertical communication techniques to this case.

In this chapter, we present the design and development of a new Pattern Time Delay Shift Coding (PDS) scheme applied to underwater acoustic communication system in shallow water environments. Simulation results for a 2kHz bandwidth communication system of a bit rate of 300 bit/s, and BER less than 10^{-4} are achieved in simulated environments using this coding scheme. Furthermore, the performance of the new method is presented and the experimental underwater tests are discussed to verify the theoretical concepts and validate the performance of the technique.

2. CODING IN UNDERWATER ACOUSTIC COMMUNICATION SYSTEMS

In this section we present some of the theoretical concepts of the Pattern Time Delay Shift Coding (PDS) technique.

The major challenge of hydro acoustic communication in underwater horizontal acoustic channels is the multipath interference. There are two aspects of the multipath interference presented in the underwater acoustic channel: code's multipath propagation time delay spread and received wave pattern distortion by multipath interference.

The model of the impulse response function of the coherent multipath channel is given by:

$$h(\tau) = A_0 \delta(\tau - \tau_0) + \sum_{i=1}^{N-1} A_i \delta(\tau - \tau_i) \quad (1)$$

Where A_i and τ_i are sound ray parameters through the receiver corresponding to the amplitude and arrival delay time respectively.

The sound ray determining the impulse response function is named acoustic Eigenray.

If signal transmitted from the sound source is denoted by z(t), the received signal in the multipath channel is given by:

$$S(t) = A_0 Z(t - \tau_0) + \sum_{i=1}^{N-1} A_i Z(t - \tau_i) + n(t) \quad (2)$$

Where the first term on the right is the direct wave, the second are refracted and reflected with boundaries. When the signals of multipath arrivals are overlapped with the direct arrival at the same time, it will derive interference, so the waveform of composite signal is different from the transmitted signal. The third term is noise; including environment noise and ship self noise.

When the delay time difference between the multipath ray and the direct ray is larger than the code's duration of a symbol, the adjacent codes (symbols) will be overlapped and interfered to cause "intersymbol multipath interference". When the delay time difference is smaller than the code's delay time resolution, "inner symbol multipath interference" will be introduced.

The simplest method to overcome the intersymbol multipath interference is to introduce enough rest-periods between transmitting of adjacent symbols, that is, the symbol interval T_0 is larger than the multipath delay spread T_d. However, such method results in a low communication rate.

The other approach is to use Frequency Shift Keying (FSK) method. In this approach the relationship between the numbers of used frequency points and the code's duration is typical given by:

$$M = \left\| \frac{T_d}{T_0} \right\| + 1 \tag{3}$$

Where $\|.\|$ is round-off operator, T_0 is the code's duration, T_d is maximum delay spread.

At the receiver system, M filters are used to separate the symbols. In the time interval of T_0, every filter only outputs the code's direct wave and multipath arrivals of same corresponding frequency, so the intersymbol interference is overcome. Although, the FSK is a well known reliable communication system used for multipath channel conditions, but it still requires broad bandwidth especially for the high bit rate digital communication conditions.

The PDS separates symbol with some pattern codes. At the receiver end, copy-correlators are used to separate the codes. To every pattern code, the copy-correlator's reference signals are different. The copy-correlator outputs a large correlation peak only when the received wave pattern is the same with reference wave pattern. If the cross correlation coefficients between every two codes are low enough, the copy-correlator can choose the corresponding code. In the time interval of T_0, the copy-correlator outputs the peak of one pattern code's direct wave and multipath waves, so the intersymbol interference is overcome.

In PDS system design, the key to suppress intersymbol multipath interference is to design a set of pattern codes, the cross correlation coefficients of which are low enough. If there are M kind of pattern codes, the delay time spread can be overcome is T_d, when $T_d = MT_0$. The PDS system uses pattern to separate codes, so it uses narrower frequency bandwidth, and it is more reliable than the FSK system. The PDS system codes with the delay time difference between the information code and its correcting code.

The multipath interference problem can be mitigated by well-known channel compensation technique such adaptive channel equalization technique. Such techniques are effective to overcome the intersymbol interference and inner symbol interference to a certain extent. Their recovery of the transmitted signal from the received signal is via estimates of the channel's impulse response. However, this will decrease the influence of multipath interference and decode more correctly. Hence, neither of the above methods is effective to cancel the effects of inner symbol interference with extremely small delay time difference (multipath opposite-phase interference, i.e., the direct wave is π phase shift to the reflected wave).

The general PDS signal frame structure is shown in Fig.1.

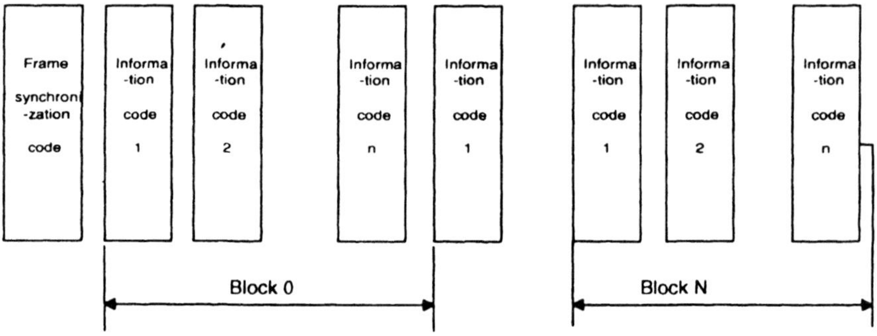

Fig.1 PDS signal frame Structure

The first code is named frame synchronization code, which provides time reference for decoding. The first block codes "Block 0" are correcting codes that provide time delay reference and detection threshold reference for corresponding pattern information recoding. The correcting codes may suppress the measuring error of time delay difference due to the inner symbol multipath interference. It's noted that the correcting codes have the same pattern with their corresponding information codes, so time delay deviation due to inner symbol multipath interference in the slowly varying channel is the same. If the correcting codes are used as time delay reference, the inner symbol multipath interference has little influence to measuring the time delay difference. The block codes after 'Block 0" are the information codes used for transmitting data or messages. The code pulse width is T_c, the time width used for coding information is T_s, and the code interval is $T_0 = T_c + T_s$, the time delay difference between information code and its reference code represents the messages. The number of pattern of frequency modulated (FM) codes is M in every block. The cross correlation coefficients between every two-pattern codes must be low enough. Their normalization cross correlation coefficients must be less than 0.35 at least.

The general block diagram of the PDS digital communication system is shown in Fig.2.

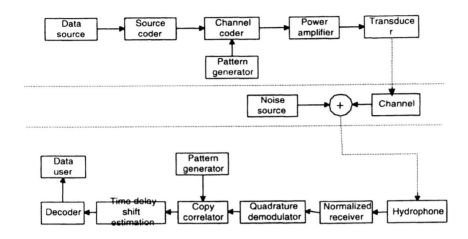

Fig.2 Block diagram of PDS underwater Acoustic communication system

The source coder produces data sequence. The channel coder modulates them into PDS time delay shift codes relatively to their correcting codes. The received signal is distorted by multipath interference and noise. Before the beginning of received message, copy-correlator searches the frame synchronization signal. Once the FM signal is detected, the copy-correlator opens a time window immediately and produces the corresponding pattern reference signal in every time window. The copy-correlation detects the code's correlation peak from the noise background. The detected code has the same pattern with the reference signal, so the time delay difference between the detected code's correlation peak and its correcting code's correlation peak can be precisely measured. "Block 0" provides the time delay difference reference for every pattern code. The information codes' time delay shift sequence is translated into data sequence or messages. For every pattern code of "block 0", the amplitudes of output peak of the copy-correlator are different. They are time-varying and influenced by inner symbol multipath interference and the channel's propagation loss. That the pattern codes' detection threshold value can be determined by the corresponding correcting codes' correlation peak multiplying a constant, which is smaller than one, it is named adaptive threshold. The adaptive threshold can suppress intersymbol interference and noise. Because every two pattern codes' cross correlation coefficient is low enough, the adaptive threshold can suppress the intersymbol multipath interference.

In brief, if the pattern codes' cross correlation coefficient is low enough, the adaptive threshold can suppress intersymbol multipath interference. Taking the correcting code's correlation peak as the reference of time delay difference can decrease the measuring deviation caused by the inner symbol multipath interference, so it can suppress the inner symbol multipath interference. To transmit one block of correcting codes during the channel coherent time interval may match the PDS codes with the slowly time-varying channel, that is, the adaptive threshold and time delay reference may track the channel's time-varying pattern.

3. TIME DELAY DIFFERENCE ESTIMATION

Since the PDS communication system uses time delay shift as information coding, hence the time delay difference estimation represent one of the key techniques in the decoding process of the system. The higher is the estimation precision required, the higher would be the communication rate.
This process may be derived from the following formula:
The PDS system communication rate is given by [22]:

$$BS = \frac{1}{T_c + T_s} \left\| \log_2 \left(\frac{T_s}{\Delta} + 1 \right) \right\| \quad \text{bit/s} \quad (4)$$

Where Δ is the quantization unit of time delay shift, T_s is the time width for information coding, $(T_c + T_s) = T_0$ is the code interval. In generally, we take

$$\Delta = \eta \delta \quad (5)$$

Where η is a constant, δ is standard deviation of time delay estimator.

To transmit data with low error rate η must be big enough. We usually take $\eta = 4 \sim 5$.

The correlator is the best time delay estimator in a band limited white noise background. In order to make the correlator's time delay resolution be higher than sampling period, interpolation technique must be adopted. In this case the "cosine interpolation" method is used.

Assume the correlation function $R(\tau)$ around the peak is an approximate cosine function, then:

$$R(\tau) = A\cos\omega\tau + B\sin\omega\tau = \sqrt{A^2 + B^2}\cos(\omega\tau - \varphi) = E\cos(\omega\tau - \varphi) \quad (6)$$

$$\varphi = tg^{-1} B/A \quad (7)$$

There are three unknown quantities in above formulas: amplitude E (correlation peak value), oscillation frequency of correlation function ω and initial phase ϕ. If three sample values around the correlation peak $R(k-1)$, $R(k)$, $R(k+1)$ are given, three equations can be derived form the formula (6). From them, ω and ϕ can be solved. The cosine interpolation time delay formulas are given by:

$$\omega = \frac{1}{\tau_s} \cos^{-1} \frac{R(k-1)+R(k+1)}{2R(k)} \tag{8}$$

$$\varphi = tg^{-1} \frac{R(k-1)\cos\omega k\tau_s - R(k)\cos(k-1)\tau_s}{R(k)\sin\omega(k-1)\tau_s - R(k-1)\sin\omega(k+1)\tau_s} \tag{9}$$

$$\hat{\tau} = k\tau_s + \frac{\varphi}{\omega} \tag{10}$$

Where τ_s is the sampling period, k is the sampling number of the discrete correlation function peak. $\hat{\tau}$ is the estimated time delay.

Correlator estimation precision of the cosine interpolation is tested by simulation experiments. The experiment condition is: pulse width of FM signal is 10ms, center frequency is 5kHz, frequency bandwidth B=2kHz, signal-noise ratio (SNR) is 5dB. The time delay estimation variances from 2000 times independent experiments list in table1. A set of deviation samples is shown in Fig.3. The time delay variance is about 8.9μs. The maximum deviation in Fig.3 is less than 34μs. When η in formula (5) is selected with values of 4 and 5(Δ=35.6μs and 44.5μs), so that the recoding error rate might be low enough.

Table- 1 below shows the results of time delay estimation variances with different signal-noise ratio with the following parameters:
T_s=10ms, f_0=5kHz, B=2kHz,

SNR(dB)	Time delay estimation variances (s^2)	Time delay estimation Standard deviation(s)
23	1.2842×10^{-12}	1.133×10^{-6}
20	2.5979×10^{-12}	1.612×10^{-6}
17	5.1506×10^{-12}	2.269×10^{-6}
14	1.0029×10^{-11}	3.166×10^{-6}
11	1.9015×10^{-11}	4.361×10^{-6}
8	3.5932×10^{-11}	5.994×10^{-6}
5	7.9533×10^{-11}	8.917×10^{-6}
2	1.7421×10^{-10}	1.320×10^{-5}

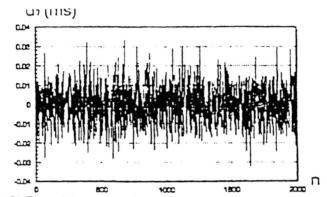
fig.(3) Time delay estimation results for 2000 times in dependent experiments
(cosine interpolation time delay estimation simulation result)

In a typical multipath channel, either noise or inner symbol multipath interference can introduce time delay estimation deviation. The codes' time delay resolution is equal to the correlation peak width which is about 1/B. If there is inner symbol multipath interference, the correlation peak waveform may be distorted by multipath interference, and the peak position and amplitude may change (see Figs.7, 9 and 13). The correlation peak may separate into two or many peaks. The constructive multipath interference (same-phased overlap with the direct wave and reflected wave) may enlarge the correlation peak, while the destructive multipath interference (opposite-phase overlap) may decrease the correlation peak. In other words, in a multipath channel, time delay measure is not definite, but time delay shift measure may achieve high precision. On the account of above, a block of correcting codes are set up in PDS system, which provide corresponding time delay shift reference for corresponding pattern information codes. For the correcting code and its corresponding information code, which have the same pattern wave, the influence caused by multipath interference is the same, so multipath interference couldn't influent time delay difference estimation precision seriously. While the adaptive threshold could track the channel's varying and provides the reasonable threshold reference for every pattern information code. Furthermore, it is well known that the channel in ocean environment is a slowly time-varying coherent multipath channel.

Hence, one block of correcting codes is needed to transmit during the channel coherent time length. In generally, it is only needed to transmit the correcting codes one times in a few ten seconds if the relative speed between the transmitter and receiver is not high. The other factor to influence the time delay estimation is the relative motion between the transmitter and the receiver. Motion can fasten the channel's varying, so correcting code transmitting must increase correspondingly. Doppler effect due to relative motion will introduce time delay measuring deviation. LFM signal's ambiguity function is coupled to Doppler and time delay. For different value of Doppler, the time delay corresponding to the correlation peak position is different. Because the correcting code and the information code have the same pattern, their time delay estimation deviations caused by Doppler are the same and the time delay difference deviation can be eliminated.

4. UNDERWATER ACOUSTIC PDS COMMUNICATION MODEL

A simulated model of the block diagram of the PDS communication system shown Fig.2 is developed to emulate the performance of the PDS method. The acoustic communication model consists of the coding modular, the decoding modular and the channel model. The typical response output of the acoustic simulated channel model is shown in Fig.4.

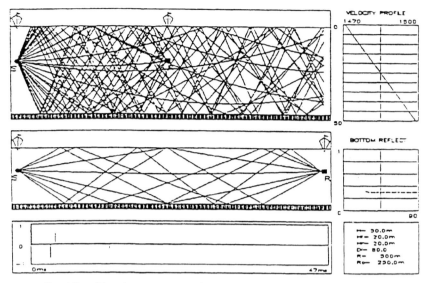

fig. (4)   Simulation result of a coherent channel

In the Fig. above the upper window on the left displays a ray trace diagram which is a set of ray trace transmitted from the source with equalized angle interval to understand the acoustic field's basic characteristic. The eigenray cluster is shown in the middle window on the left. The eigenrays are a series of special sound ray passing through the receiver which the parameters (A_i, τ_i,) determine the channel's impulse response function are shown in the lower window on the left of the Fig. The calculation error of time delay of all eigenray is less than 3μs. The number of egienray is related to the following effects: transducer's directivity, geometric position of the transmitter and receiver, sea depth and profile of sound velocity etc. Once the channel impulse response function is calculated, the simulation program can predict the received signals form for the transmitted source. The sound velocity profile is shown in the upper window on the right. The simulation program enables to input actual measuring sound velocity profile, so communication quality in any condition could be observed. The middle window on the right displays bottom reflection coefficient curve. The simulation program enables to input arbitrary bottom reflection characteristics. The simulation parameters are listed in the lower window on the right. All parameters could be changed to analyze the communication quality. Some of the simulation results of this simulated model are shown in Fig.5 and Fig.6 respectively.

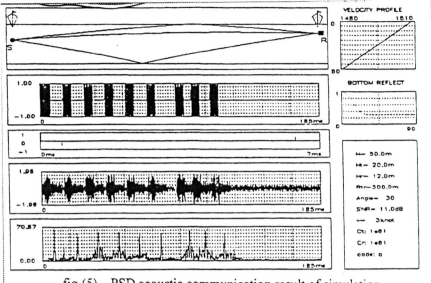

fig.(5) PSD acoustic communication result of simulation
(inner symbol destructive multipath interference)

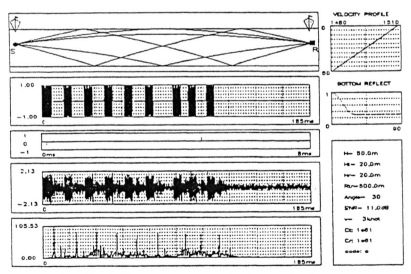

fig. (6) PSD acoustic communication result of simulation
(inner symbol constructive multipath interference)

In these Figs. (from top to bottom), the left windows show eigenray diagram, transmitted PDS code signal, received wave disturbed by noise and multipath interference and output signal form the copy-correlator respectively. while the right window represent the sound velocity profile, bottom reflection coefficient and main parameters of the simulation system.

Fig.5 and Fig.6 show the simulation results of inner symbol multipath interference. The former considers the inner symbol destructive multipath interference, while the later considers the inner symbol constructive multipath interference. On the account of above, the former received signal is not more regular than the later, and its amplitude is smaller. The copy-correlator output peak amplitude of the former is smaller, and its correlation peak waveform is distorted as shown in Fig.7. The simulation results of above channel conditions are listed in the Table 2 and Table 3 respectively. It can be concluded from these results that in the inner symbol destructive multipath interference case, the time delay estimation deviation is increased considerably with a corresponding error rate increase. In order to reach high reliable communication effect in such channel, higher SNR is needed.

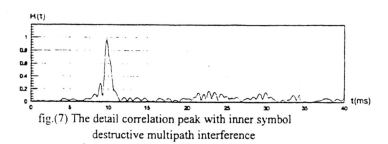

fig.(7) The detail correlation peak with inner symbol destructive multipath interference

Table 2 Time delay estimation deviation and error rate in inner symbol destructive multipath interference (simulation results)

Rate SRN	292 bit/s		383 bit/s		453 bit/s	
	σ_t(ms)	Pe	σ_t(ms)	Pe	σ_t(ms)	Pe
17dB	0.01769	0	0.01775	0	0.01747	0.00053
11dB	0.02999	0	0.03047	0.00022	0.03061	0.0074
7.4dB	0.03952	0.00022	0.04102	0.0023	0.04046	0.0194

Table 3 Time delay estimation deviation and error rate in inner symbol constructive multipath interference (simulation results)

Rate SRN	292 bit/s		383 bit/s		453 bit/s	
	σ_t(ms)	Pe	σ_t(ms)	Pe	σ_t(ms)	Pe
17dB	0.003248	0	0.003113	0	0.003182	0

| 11dB | 0.005986 | 0 | 0.006134 | 0 | 0.006178 | 0.00017 |
| 7.4dB | 0.009043 | 0 | 0.009084 | 0 | 0.009149 | 0.00024 |

5. EXPERIMENTAL RESULTS OF THE SYSTEM

In this section the experimental set up and the underwater test results of the system are presented. The aim of these underwater tests is to evaluate the performance of the PDS communication system and to study the influence of the tested underwater channel on the performance of the system. The block diagram of the experimental test configuration is shown in Fig.8.

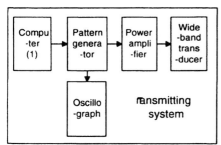
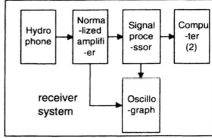

Fig.8 Block Diagram of the Experimental system

The underwater experimental test system is consisting of transmitter system and receiver sub-systems. The transmitter sub-system consist of the transmitter DSP-based computer system I, pattern generator, power amplifier and wide-band transducer (with non-directivity). The pattern generator is composed of (TMS320C25) fixed point DSP system. The computer system is used for source coding, channel coding and controlling the pattern generator to transmit PDS codes. The electric power of the power amplifier is about 30W. There are 5 pattern codes with center frequency (f_0) of 5kHz, band width of 2kHz, frame synchronization code duration of 20ms, information coding time length $T_s=10$ms, duration of information code pulse is 10 ms, sampling frequency $f_s=6f_0$. The communication bit rate is changed from 70 bit/s and 300 bit/s respectively.

The receiver system includes hydrophone (non-directivity), normalized amplifier and filter, signal processor (TMS320C25) and computer II. The signal processor calculates copy-correlation estimates time delay difference and decode in real time. At the same time, raw data and correlation output is transmitted from the signal processor to the computer II. The raw data and output data are collected and displayed in computer II in real time. The copy-correlation output also may be observed through the oscilloscope after D/A conversion.

The reservoir trial is performed in a wide water area with the depth of 40m at the SongHua Lake in JiLin province in China. The transmitting ship is fixed beside an island in the middle of the lake, while the receiver ship drifts with the speed of about 3 knot. Surface wave height is about 30cm. The transducer and the hydrophone are dipped with the depth of 15m and 5m respectively. The system communicates data over a horizontal range of 1500m. With the drifting of the receiver ship, the system marks statistic error rate, collects data, displays correlation output and receives signal waveform and decodes continuously during the drift. The correlation output waveforms at different distances are shown in Figs. 9-11. The maximum error rate obtained with three trials of the setup was less than 10^{-4}.

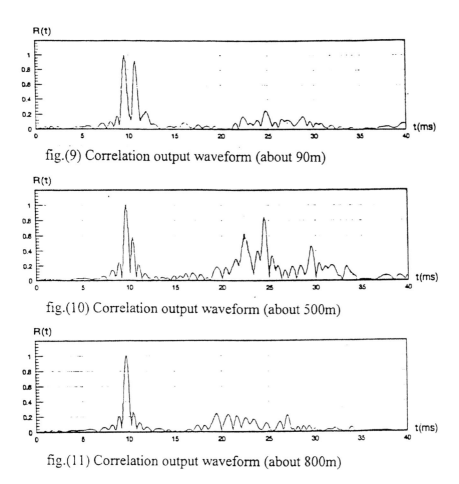

fig.(9) Correlation output waveform (about 90m)

fig.(10) Correlation output waveform (about 500m)

fig.(11) Correlation output waveform (about 800m)

The Figs above show the correlation peak of the frame synchronization code varying with the distances tested. The distances (270m, 460m) with high correlation peak as shown in Fig.12 correspond to the case of inner symbol constructive multipath interference. For the test distance (300m) with low correlation peak as shown also in Fig.12 correspond to the case of the inner symbol destructive multipath interference. In the condition of destructive multipath interference the maximum error rate were at the level of 10^{-4}.

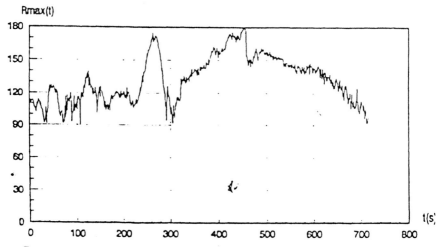

fig. (12) Correlation output peak varying with time (distance)
(relative speed is about 3 knot, experiment results)

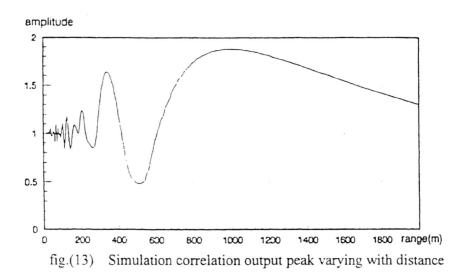

fig.(13) Simulation correlation output peak varying with distance

The results shown in Fig.13 confirm that that the simulation results of the system's model correlate well with the corresponding results of the actual underwater trial conditions. Fig.14 and Fig.15 give the trial's correlation output waveform every second during a testing period of 10s. This illustrates that the channel is stable enough even if the receiver ship drifts at speed of 3 knot.

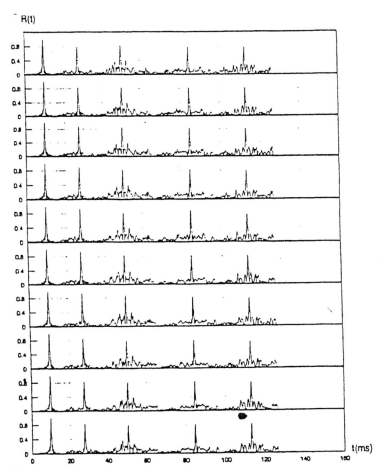

fig (14) Correlator output wave group in every 1s

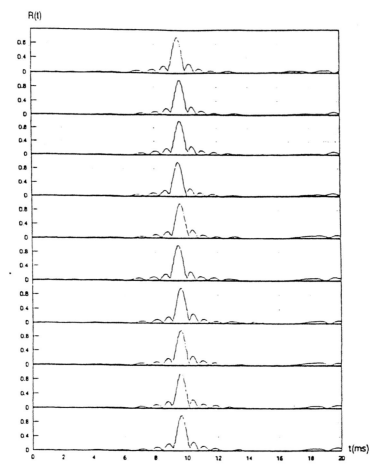

fig.(15) Detail correlator output wave of fig.(14)

6. CONCLUSIONS

High-speed underwater acoustic digital communication is constrained by the channel multipath interference whose influence is presented in the inner symbol multipath interference and the intersymbol multipath interference. The later limits the communication rate, while the former influents the communication reliability and error rate. To overcome these effects, a typical FSK communication that divides the frequency band to circumvent the intersymbol multipath interference and uses a group of filters to separate codes with different frequency is used. However, this methodology requires wide bandwidth for high data rate transmission requirements.

In this chapter a new PDS communication system technique that separates codes by a group of copy-correlators with different reference signals is presented. Hence, it can circumvent the intersymbol multipath interference to achieve high data rate with relatively narrow bandwidth. The key features of this method are that the cross correlation coefficients between every two-pattern code must be low enough and their relevant normalization cross correlation coefficients are less than 0.35.

A simulation system and real underwater test results are presented to evaluate and correlate the performance of the PDS communication system. This study illustrates that PDS communication system provide a more robust approach especially in shallow water channel conditions. In the underwater trials, the system was able to transmits data with bandwidth of 2kHz and bit rate of 300 bit /s with a maximum error rate around 10^{-4}.

Further work on using relevant channel equalization techniques can improve further the performance of the system in terms of better BER and higher transmission rates and this is could be a further development to the method presented in this work.

7. REFERENCES

1. Dr.Rodney Coates, "Underwater Acoustic Communication", Sea Technology, pp41-47, July, 1994.
2. L.Wu and A.Zielinski, "Multipath rejection using Narrow Beam Acoustic Link", IEEE,pp287-290, 1988.
3. Manora K.Caldera, "A Multi-frequency Digital Communication Technique for Acoustic Channel with Multipaths" IEEE, pp140-145, 1987.
4. Michiya Suzuk, et al, "Digital Acoustic Telemety of color Video Information", Proc. Ocans'89, Settle, Washington, 1989, pp892-896.
5. P. Bragard, G. Jourdain, and J. Martin, "Optimal Adaptive Algorithms Behavior used in underwater Communication Signal Equalization", Signal processing IV:Theories and applications, pp363-366, 1988.
6. Milica Stojanovic, Josko A. Catipovic, John G.Proakis, "Phase Coherent Digital Communication for Underwater Acoustic Channels", IEEE Journal of Oceanic Engineering, Vol. 19, No.1,pp100-111, January, 1994.
7. Qian Wen and James A. Ritcey,"Spatial Diversity Equalization Applied to Underwater Communications", IEEE Journal of Oceanic Engineering, Vol.19, No.2, pp227-241, April 1994.
8. J.S.Collins, J. R. Galloway, and M. R. Balderson,"Auto Aligning System for Narrow Band Acoustic Telemetry", Occans'85, San Diego.

9. H.Davidson, etal, "Reliable Underwater Acoustic Data Link Employing an Adaptive Receiving Array",Proc. of the inst. of Acoustics, Vol.9, April, 1987.

10. K.Akio, et al,"An Acoustic Communication System for Subsea Robot", Proc.Ocans'89 Settle, Washington,1989,pp765-770.

11. Wen Zhoubin, Feng Haihong, Hui Junjing "A New Uderwater Acoustic Communication System", Journal of Acoustics(Chinese) Vol.18, No.5, Sept., 1993,pp892-896.

12. O.R.Hinton, G.S. Howe, and Adams," An Adaptive High Bit Rate Subsea Communication System", European Conference on Underwater Acoustics, 1992,9, pp75-79.

13. Arthur B. Baggeroer, "Acoustic Telemetry-An Overview", IEEE Journal of Oceanic Engineering, Vol. ,OE-9, No.4, pp229-235, October, 1984.

14. M. Stojanovic, J.Catipovic and J.G.Proakis, "Adaptive Multichannel Combining and Equalization for Underwater Acoustic Communication", J.Acoust. Soc.Am., Vol.94, No.3, Sept.,pp1621-1631, 1993.

15. P.Bragard and G.Jourdain, "Adaptive Equalization for Underwater Data Transmission", IEEE, pp1171-1174, 1989.

16. G.Hafizimana, G.Jourdain, and G. Loubct, " Codind for Communication through Multipath Channels and Application to Underwater Casc", Signal Processing III: Thcorics and Applications, pp 1087-1090, 1986.

17. B. Solaiman, A. Glavicux, and A. Hillion, "Performance of Slow Frequency Hopping BPSK System Using Convolutional Coding in Underwater Acoustic Media", Proc. Icassp'88.

18. Josko A.Catipovic, "Performance Limitations in Underwater Acoustic Telemetry", IEEE Journal of Oceanic Engineering Vol.15 No.3, pp 205-216, July, 1990.

19. P.Bragard, G. Jourdain, and Martin."Optimal Adaptive Algorithm Behavior Used in Underwater Communication Signal Equalization", Signal Processing IV:Theories and Application, 1988, pp363-366.

20. Salvatore D. Morgera, Keith A. Reuben, and Cedric Cole, "A Microprocessor-Based Acoustics Telemetry System for Tide Measurement", IEEE journal of Oceanic Engineering, Vol. Oe11, No.1, January, 1986, pp 100-108.

21. G.H. Sandsmark, "High Speed Underwater Acoustic Data Transmission: A Brief Review", Proc. of Int. Workshop on Marine Acoustics ,pp133-140, Mar, 1990, Beijing.

22. Hui Junjing. "Underwater Acoustic Channel"(Chinese), Defence Industry Press, 1991.

8

Three-dimensional underwater acoustical imaging and processing

Andrea Trucco[1], Maria Palmese[1], Andrea Fusiello[2], and Vittorio Murino[2]
[1] Department of Biophysical and Electronic Engineering, University of Genova, Italy
[2] Department of Computer Science, University of Verona, Italy

1. INTRODUCTION

Acoustic imaging is an active research field that aims to study techniques for the formation and processing of images generated by raw signals acquired by an acoustic system [1]. Our purpose is to present a brief survey concerning the generation and processing of acoustic images for underwater applications [2,3], especially focusing on algorithms for three-dimensional (3-D) imaging. Like optical systems, acoustic systems can generate an image by processing the waves backscattered from the objects of a scene. The relative ease of measuring the time-of-flight of an acoustic signal makes it possible to generate not only acoustic 2-D images similar to optical ones but also range estimates that can be used to produce a real 3-D map.

The main advantage of acoustic imaging systems over optical ones is that they work properly also beyond the optical visibility range. Underwater optical vision provides images with finer resolution, but its range is limited to a few tenths of meters in very good water conditions, and the situation is even worse when mud in suspension raised by underwater tasks sharply reduces visibility [2]. Although larger than the range of optical visibility, the range of applicability of a 3-D acoustic imaging system varies according to the specific sensor adopted and the signal frequency characterizing the sensor. Generally, high frequencies are utilized (from about a hundred kilohertz to few megahertz) over a range going from few centimeters to a hundred meters.

In general, the scene under investigation is first insonified by an acoustic signal, then the backscattered echoes acquired by the system are processed to create an image of the scene. This process can be performed by two different approaches [3]: use of an acoustic lens followed by a retina of acoustic sensors, or acquisition of echoes by a two-dimensional (2-D) array of sensors and subsequent processing by adequate algorithms, thus avoiding the need for a physical lens. Such algorithms belong to the beamforming or the holography class.

To date, unlike the applications of two-dimensional imaging, the practical utilization of 3-D acoustic imaging is limited because of several scientific and technological issues that make it difficult and expensive to produce imaging systems working in real time. However, some recent achievements, in both hardware and software areas, allowed the production of a few prototypes and commercially available systems.

Once an acoustic image is generated, several image processing algorithms can be designed for the detection and reconstruction of the objects contained in the observed scene. Depending on the type of the generated image, i.e., whether it embeds three-dimensional (3-D) information together with intensity data or not, several approaches can be followed for the image segmentation. Actually, image filtering and segmentation are the first and fundamental phases after the generation of an image, and constitute the base for any subsequent processing and interpretation, and segmentation and reconstruction are natural subsequent steps. In this chapter, such approaches will be surveyed, especially those devoted to 3-D image processing, leaving aside other methods specifically devoted to high-level image interpretation. Finally, a complete system for segmentation and reconstruction of underwater tubular structures is described, also showing the related high-level applications.

In this chapter, a model of the interaction of the acoustic energy with the scene to be imaged is presented in Section 2. In Section 3, specific methods (i.e., beamforming, holographic, and lens-based techniques) are outlined, focusing attention on the 3-D beamforming approach. In Section 4, a real system that exploits a matrix approach based on the holographic method is detailed, and in Section 5, the arrangement and representation of a 3-D image are described.

Section 6 deals with the state of the art literature about the 3-D image processing, focusing especially on acoustic 3-D image segmentation and reconstruction. A specific method for the segmentation and reconstruction of 3-D acoustic images is detailed in Section 7. In this part, a complete image processing system is presented, focusing on several segmentation algorithms, and showing possible applications which inspired this work. Finally, conclusions are drawn in Section 8.

2. DATA MODEL

In underwater 3-D imaging, a typical scene is composed of several solid and continuous objects. One of the most powerful methods for computing the field returned by a complex and realistic underwater object is to represent its surface by a collection of densely packed point scatterers or small facets [4,5,6].

We assume that the imaged scene is made up of M point scatterers; the i-th scatterer is placed at the position $\mathbf{r}_i = (x_i, y_i, z_i)$ and its distance from the coordinate origin is equal to $r_i = |\mathbf{r}_i|$, as shown in Fig. 1a. We can define the plane $z = 0$ as the plane that receives the backscattered field (see Fig. 1a). If an acoustic pulse $q(t)$ is emitted by an ideal point source placed in the coordinate origin, and if we assume that a spherical propagation occurs inside an isotropic, linear, absorbing medium, then the Fourier transform of the pressure measured at the position $\mathbf{p} = (x, y, 0)$ and due to the action of the M point scatterers [3] is equal to:

$$S(\omega, \mathbf{p}) = \sum_{i=1}^{M} D_i \frac{\omega^2 e^{-2\alpha(\omega) r_i}}{c^2 r_i^2} Q(\omega) e^{-j\frac{\omega}{c}(r_i + |\mathbf{p} - \mathbf{r}_i|)} \tag{1}$$

where $Q(\omega)$ is the Fourier transform of the emitted pulse $q(t)$, $\alpha(\omega)$ is the absorption coefficient of water, $\omega = 2\pi f$ is the angular frequency, f is the frequency, c is the sound velocity in the medium, and D_i is a coefficient dependent on the radius of the i-th scatterer and on the densities and the compressibilities of the propagating medium and of the i-th scatterer [3,5].

For typical frequencies and distances involved in 3-D underwater imaging, it has been shown [3] that the factor ω^2 in (1) is compensated for by water absorption. Therefore, we can introduce a constant factor σ_i (dependent on the distance r_i) to replace the term $\omega^2 \exp\{-2\alpha(\omega) r_i\}$, and we can rewrite equation (1) as:

$$S(\omega, \mathbf{p}) \cong \sum_{i=1}^{M} C_i \, Q(\omega) e^{-j\frac{\omega}{c}(r_i + |\mathbf{p} - \mathbf{r}_i|)} \tag{2}$$

$$C_i = D_i \frac{\sigma_i}{c^2 r_i^2}. \tag{3}$$

Moreover, if the cone whose vertex is in the coordinate origin and that contains the scene volume to be imaged has an angular aperture that is not too wide, we can use the Fresnel approximation for the Green function [7], [8] to write the term $|\mathbf{p} - \mathbf{r}_i|$ in the phase of (2) as follows:

$$|\mathbf{p} - \mathbf{r}_i| \approx r_i - \hat{\mathbf{r}}_i \cdot \mathbf{p} + \frac{p^2}{2r_i} \tag{4}$$

where p is the modulus of $\mathbf{p}$, and $\hat{\mathbf{r}}_i$ is a unitary vector equal to $\mathbf{r}_i / r_i$.

Finally, if we restrict our reasoning to the plane $y = 0$ (see Fig. 1b), then $\mathbf{p} = (x, 0, 0)$, $\mathbf{r}_i = r_i(\sin\beta_i, 0, \cos\beta_i)$, β_i being the angle between the vector $\mathbf{r}_i$ and the z-axis, and (2) can be simplified as follows:

$$S(\omega, x) = \sum_{i=1}^{M} C_i \, Q(\omega) e^{-j\frac{\omega}{c}\left(2r_i - x\sin\beta_i + \frac{x^2}{2r_i}\right)}. \tag{5}$$

The angle β_i refers to the arrival angle as it indicates the direction of the echo of the i-th scatterer.

The approximation for the term $|\mathbf{p} - \mathbf{r}_i|$ that has been discussed is of great practical importance as dealing with all combinations of $\mathbf{p}$ and $\mathbf{r}_i$ is sometimes very difficult (even in modern imaging systems exploiting digital technology).

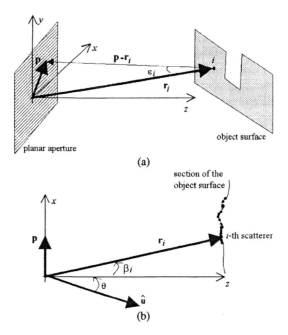

Figure 1. Notation and geometry of the data model. (a) 3-D representation. (b) 2-D projection on the plane $y = 0$.

3. ACQUISITION OF 3-D INFORMATION

In 3-D underwater imaging, a scene is typically illuminated by the emission of an acoustic pulse, and the backscattered echoes are collected over a 2-D aperture and processed to create an image of the scene. We can choose between two different approaches [3,1] to perform the operation of echo processing: use of acoustic lens followed by a retina of acoustic sensors, or acquisition of the echoes impinging on a planar array of sensors and processing of such echoes by appropriate algorithms. Beamforming and holographic methods are two processing approaches that can be successfully exploited in 3-D imaging systems.

Acoustic lenses work like optical ones: backscattered echoes are focused on an image plane where a 2-D retina of sensors transforms the acoustic image into electrical signals. Thanks to the facility of measuring the time-of-flight of an acoustic pulse, one can generate range estimates that can be utilized to produce a real 3-D map. Each sensor of the retina placed behind a lens receives a signal that represents the scene response coming from a well-defined direction. By collecting the signals of all the retina sensors, one can obtain complete information about the 3-D structure of the scene.

Beamforming systems collect backscattered echoes by a 2-D array of sensors only once; then, they arrange the echoes in such a way as to amplify the signal coming from a fixed direction (steering direction) and to reduce all the signals coming from any other directions. As the output signal gives information about the scene structure in the steering direction, it is possible to create a 3-D image by repeating the beamforming process after fixing many adjacent steering directions, as in a raster-scan operation.

Holographic systems also start from the echoes acquired by a 2-D array of sensors, but they aim to reconstruct the 3-D structure of a scene by back-propagating the received signals. Acoustic holography is a special case of inverse diffraction and is performed through the inversion of the propagation and scattering equations. An image is not generated by a raster-scan operation, but the holographic algorithm produces the whole image at the same time.

In the rest of this work, we focus on the beamforming approach and, in a less detailed way, on the holographic approach.

3.1 Beamforming

Beamforming is a spatial filter that linearly combines the temporal signals spatially sampled by a discrete antenna, i.e., an array of sensors placed according to a known geometry.

Let us consider a set of N point-like and omnidirectional sensors that constitute a receiving 2-D array, numbered by the index l, from 0 to $N-1$. Denoting by $\mathbf{p}_l$ the position of a given sensor of the set on the plane $z = 0$ and by $s_l(t)$ the signal received by the sensor and linearly proportional to the pressure field, one can compute the beam signal [9], $b(t, \hat{\mathbf{u}})$, steered in the direction of the unitary vector $\hat{\mathbf{u}}$ by using the following definition:

$$b(t, \hat{\mathbf{u}}) = \sum_{l=0}^{N-1} \varphi_l \, s_l(t - \tau(\hat{\mathbf{u}}, r_0, l)) \tag{6}$$

$$\tau(\hat{\mathbf{u}}, r_0, l) = \frac{r_0 - |\mathbf{p}_l - r_0 \hat{\mathbf{u}}|}{c} = \frac{r_0 - \sqrt{r_0^2 + |\mathbf{p}_l|^2 - 2r_0 \hat{\mathbf{u}} \cdot \mathbf{p}_l}}{c} \tag{7}$$

where φ_l are the weights assigned to each sensor and r_0 is the focusing distance. The net result is the formation of a temporal signal in which the contributions coming from the direction $\hat{\mathbf{u}}$ and the distance r_0 are amplified, whereas those coming from other directions and distances are attenuated.

Now, let us restrict our reasoning to the plane $y = 0$ and assume all the weights φ_l to be unitary; then the steering direction can be indicated by the angle θ measured with respect to the z axis (see Fig. 1b). Beginning from equations (5) and (6), by applying the Fresnel approximation to the delay in (7) and by assuming the distance of all the scatterers making up the scene to be equal to r_0, one can obtain [3] the following expression for the beam signal:

$$b(t, \theta) = \sum_{i=1}^{M} q\left(t - \frac{2r_i}{c}\right) C_i \, BP_{BMF}(\overline{\omega}, \beta_i, \theta) \cdot \tag{8}$$

$$BP_{BMF}(\omega, \beta, \theta) = \frac{\sin[\omega Nd(\sin\beta - \sin\theta)/2c]}{\sin[\omega d(\sin\beta - \sin\theta)/2c]} \tag{9}$$

where $BP_{BMF}(\omega, \beta, \theta)$ is a reception diagram commonly called beam pattern, which depends on the arrival angle β, the steering angle θ, and the angular frequency ω. We have assumed the array to be equispaced and centred in the coordinate origin, d to be the inter-element spacing, and the beam pattern to have a constant profile over the signal bandwidth centred in the angular frequency $\overline{\omega}$.

With reference to an array composed of 40 1.5 mm-spaced elements, Fig. 2 shows some beam patterns as a function of the arrival angle (visualized on a logarithmic scale normalized to 0 dB, provided that the absolute values are considered) for different frequency values and steering angles.

Equation (8) clearly shows that each scatterer contributes to the beam signal by adding a replica of the acoustic pulse $q(t)$, delayed on the basis of its distance r_i, weighted by its constant C_i (strictly related to the scatterer reflectivity) and by the beam pattern value that depends on the discrepancy between the arrival angle β_i of the scatterer and the steering angle θ. Owing to the profile of the beam pattern, the contributions of the scatterers characterized by arrival angles very close to the steering angle have predominant magnitudes.

The beam pattern of an imaging system is very useful to evaluate the system performances. A conventional beam pattern presents a main lobe in the steering direction and sidelobes of minor magnitudes in other directions. The width of the main lobe is the measure of the angular resolution (also called lateral resolution) of the imaging system, whereas the generation of artefacts degrading useful information depends on the level of the sidelobes. When the weight coefficients are unitary, the following equation [9] provides the arrival angles at which the main lobe is reduced to −3 dB, thus giving a measure of the main lobe's width:

$$\sin\beta_{-3dB} = \sin\theta \pm 0.44 \frac{\lambda}{Nd}. \tag{10}$$

To improve the angular resolution, one can increase the number of elements, the frequency, or the inter-spacing. Nevertheless, if the frequency increases, a grating lobe appears in the beam pattern (see Fig. 2d). This aliasing effect is due to the spatial under-sampling that occurs when the array elements are equispaced and the inter-element spacing is larger than $\lambda/2$. To avoid ambiguity effects resulting from the presence of grating lobes in the beam pattern, it is necessary to limit both the insonification and steering operations inside a narrower angular sector [1]. In greater detail, the maximum steering angle to avoid ambiguities is the following:

$$\theta_{max} = \pm\arcsin\frac{\lambda}{2d}. \tag{11}$$

The beam patterns have been developed and discussed for a plane space and taking into consideration a linear array, although this configuration does not allow one to obtain 3-D images but only a section of a scene profile (see Fig. 1b). To steer the beam inside a 3-D space, two steering angles should be used and a planar array is mandatory. Figure 3a shows the geometry and the notation for a planar array, where β_a and β_e are the azimuth and elevation arrival angles, respectively, and θ_a and θ_e are the azimuth and elevation steering angles, respectively. Figure 3b shows the beam pattern for an array composed of 20×20 elements that are $\lambda/2$ apart, when the steering angles are $\theta_a = 40°$ and $\theta_e = 0°$.

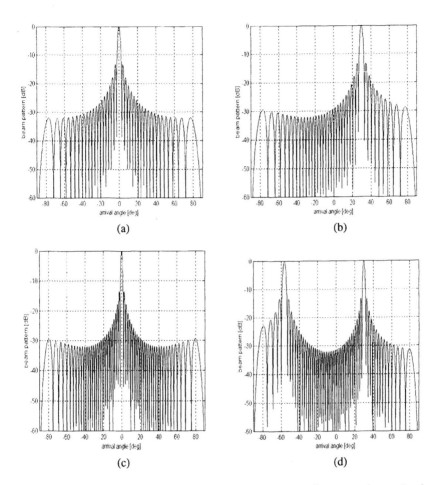

Figure 2. Beam power patterns of a 40-element array with 1.5 mm spacing and unitary weights. (a) Frequency = 500 kHz (i.e., inter-element spacing = $\lambda/2$), steering $\theta = 0°$. (b) Frequency = 500 kHz (i.e., inter-element spacing = $\lambda/2$), steering $\theta = 30°$. (c) Frequency = 750 kHz (i.e., inter-element spacing = $3\lambda/4$), steering $\theta = 0°$. (d) Frequency = 750 kHz (i.e., inter-element spacing = $3\lambda/4$), steering $\theta = 30°$.

3.2 Resolution cell

At this point, it may be useful to clarify how beam signals can be exploited to generate a 3-D image. To this end, it is necessary to define the range resolution and to recall the angular resolution. Inside a wide interval called depth of field, centred at the focusing distance, the range resolution is defined as the minimum distance between two equal scatterers (placed in the same beam direction) that is needed to resolve their responses. The range resolution [6,10] is typically inversely proportional to the bandwidth of the emitted pulse $q(t)$. Analogously, the angular resolution is the minimum angular spacing that allows two equal scatterers, placed at the same distance from the array centre, to be resolved. A resolution cell can be defined as a volume whose dimensions can be derived from the knowledge of the system's angular and range resolutions and inside which it is not possible to separate scatterer contributions. To arrange a grid of resolution cells that covers the whole volume to be imaged without leaving free holes one should plan the number of beam signals to be computed and their angular spacing and choose the sampling frequency of each beam signal in agreement with the range resolution [9].

The acoustic responses of the scatterers contained inside a cell are a function of the reflectivity and of the relative position of each scatterer; the reflectivity value to be assigned to each cell can be derived from the amplitude of the beam signal over the time interval related to such cell.

Resolution cells span the whole volume to be imaged, whereas scatterers are mainly placed on object surfaces; then, starting from this collection of cells, one can organize the effective information in more compact ways, discarding useless cells or directly extracting object surfaces.

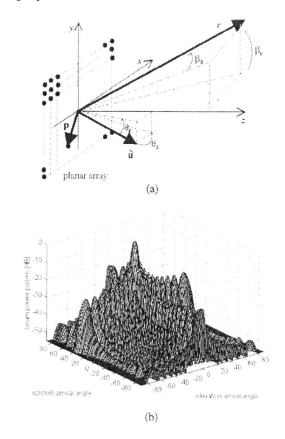

Figure 3. (a) Notation and geometry of a 2-D array. (b) Beam power patterns of an array composed of 20×20 elements that are $\lambda/2$ far from one another, when the steering angles are

$\theta_a = 40°$ and $\theta_e = 0°$ and the weights are unitary.

4. MATRIX APPROACH AND REAL SYSTEMS

The holographic approach is characterized by many possible approximations that provide different practical algorithms. Here we present a holographic method capable of avoiding the Fresnel approximation (whose validity region is often too restricted) and to process wide-band signals (very important to obtain a good range resolution). This method is based on a matrix formulation that is used by a few real 3-D systems [11,12].

To this end, it is necessary to express the data model presented in (2) in a different manner. Let us denote by $s(\omega,p)$ the $N \times 1$ column vector of the field received by N sensors placed at p_l (l being an integer ranging from 1 to N), and by $c(\omega,r)$ the $M' \times 1$ column vector of the reflectivity of each resolution cell contained in the scene volume to be imaged. The number of cells, M', is different from the number of scatterers, M, used in the previous data model, and if the i-th cell placed at r_i (i being an integer ranging from 1 to M') does not contain any object, its reflectivity may be null. The data model presented in (2) can be rewritten as follows:

$$s(\omega, p) = U(\omega, p, r) c(\omega, r) \tag{12}$$

where $U(\omega,p,r)$ is an $N \times M'$ transfer matrix whose element u_{li} is defined as:

$$u_{li} = Q(\omega) e^{-j\frac{\omega}{c}(r_i + |p_l - r_i|)} \tag{13}$$

The imaging process lies in estimating the vector c starting from the knowledge of U and the experimental measure of s.

As in a 3-D imaging system, the total number of resolution cells is too large to be handled by a single matrix, one can rewrite (12) in the following way:

$$s_n(\omega, p) = U_n(\omega, p, r) c_n(\omega, r) \tag{14}$$

where c_n is a regular grid of cells on a spherical surface of radius r_n, s_n is the Fourier transform of the field received over a brief time interval starting at $t_n = 2r_n/c$, and U_n is the related transfer matrix. As a result, the resolution cells of the 3-D volume are organized as a sequence of concentric spherical layers, and a specific transfer matrix is defined for each layer. The best estimate, $\tilde{c}_n$, of the vector c_n [11,12], assuming $M' > N$, is given by the minimum-norm solution of (14), that is:

$$\tilde{c}_n = U_n^H \left(U_n U_n^H \right)^+ s_n \tag{15}$$

where ^{H} is the complex conjugate and transpose, $^+$ is the pseudoinverse, and the matrix arguments are neglected to simplify the notation.

Among 3-D underwater imaging systems that exploit the above matrix approach the EchoScope 1600 designed and produced by OmniTech in Norway [12,13,14] is the only off-the-shelf 3-D acoustic camera available.

The number of acoustic sensors is 1600, which make up a 40×40 array with a 19.5-cm side, able to work at three different frequencies: 150, 300, and 600 kHz. The value of the angular frequency ω is fixed, thus the Fourier transform is not necessary as the amplitude and phase data on a received signal are obtained by a quadrature reception [9]. For each spherical surface of radius r_n, a grid of 64×64 partially overlapped resolution cells is determined, and the best estimate of the related vector c_n is obtained by means of (15), where the spectral theorem is exploited to decompose the matrix $(U_n\ U_n^H)$, as described in [12]. As the inter-sensor spacing is fixed, to avoid grating-lobe effects the viewing and insonification angles are bounded as follows: 90°×90° at 150 kHz, 50°×50° at 300 kHz, and 25°×25° at 600 kHz, and the lateral resolutions (measured at −3 dB, with unitary weights) are 2.5°, 1.3°, and 0.6°, respectively. The allowed detection range spans from 1 to 100 m, and, thanks to the powerful computation architecture, it is possible to achieve on average five 3-D images per second with a range resolution of 5 cm. Owing to the above-mentioned features, the EchoScope 1600 represents a powerful way of obtaining real-time 3-D information over a very wide range of distances, thus allowing one to choose the best tradeoff among resolution, viewing angle, and side-lobe level.

5. 3-D IMAGE REPRESENTATION

Independently of the imaging system used, the obtained 3-D information is organized into a dense lattice of resolution cells of different dimensions to cover the whole volume of interest. Essential information about each cell lies in the coordinates of the cell centre and in the acoustic amplitude or intensity. Sometimes, to achieve an initial cleaning of the imaged volume, a first process discards all the resolution cells whose acoustic amplitudes do not exceed a given threshold. The aim is to remove the effects of electronic noise and, in particular, of the sidelobes in the beam pattern of the acoustic system. The resolution cells can be projected onto a regular 3-D grid of voxels (volume elements) of constant dimensions in a Cartesian coordinate system. This operation is called "scan conversion." To avoid loss of resolution, each voxel should be smaller than the smallest resolution cell to allow one or more voxels to be contained in a given resolution cell. To this end, there are two possible approaches: (1) searching for the voxel containing the centre of a given resolution cell, assigning the acoustic amplitude of the cell to such a voxel, repeating this procedure for all the cells, and, finally, interpolating in order to assign a value also to each voxel that does not contain a cell centre; (2) computing how many voxels are inside a given cell on the basis of the cell dimensions, assigning the acoustic amplitude of the cell to the voxels, repeating this operation for all the cells, and, finally, checking if some voxel is unassigned. The resulting regular grid of voxels may be a useful starting point for many post-processing purposes, and also facilitates the following operations: (1) extracting a sequence of 2-D images representing the acoustic responses of successive planar slices of a scene, (2) extracting several 2-D images cut with different orientations and projecting the 3-D volume by performing an orthogonal or perspective projection integrating the acoustic amplitudes of the cells met along the projection line.

If we are interested in the first object surface that is met starting from the array centre and following the steering direction, a more compact method to exploit the information contained in the set of resolution cells is to use a couple of 2-D images, one for the amplitude and the other for range information. In this case, we expect to extract only two values from each beam: the distance of the object surface and the amplitude of the acoustic response. The distance of a scattering object can be estimated by searching for the maximum peak (or the first peak that exceeds a given threshold) of the envelope of the beam signal and, starting from the time instant at which that peak occurs, by deriving the distance from simple geometrical considerations. Therefore, for each steering direction $\hat{u}$, a triplet ($\hat{u}$, distance, amplitude) can be extracted, so 3-D data reduce to a set of triplets, the number of which is equal to the number of beam signals. Then, the collection of triplets can be projected onto the two images, according to the kind of information contained in each image.

6. ACOUSTIC 3-D IMAGE PROCESSING

A large body of literature is available concerning the segmentation of range images, but few papers deal with acoustic range image processing. In general, looking at the state of the art of three-dimensional image processing, three possible research lines can be identified for range image segmentation: edge-based, region-based, and hybrid methods.

Edge-based methods [15]-[19] aim at determining step and roof edge points using separate procedures, mainly based on the analysis of local depth discontinuity, and the discontinuity of first derivative, respectively. Due to the low resolution and high noise level typical of acoustic images, edge-based approaches are not pursued in this application field, and most of the methods are devoted to the design of region-based algorithms.

The goal of region-based methods is to determine image areas with similar differential geometrical properties. Typically, all these approaches follows the so-called *local-to-global* scheme composed by two phases. A pre-processing filter is preliminary applied to reduce noise effects. In the first *local* phase, an over-segmentation is first determined, labelling each point on the basis of the local differential features. The second *global* phase consists in a fusion stage, where regions detected in the 1st phase are merged if leading to a better segmentation. In practice, the first important step lies in the choice of the type of surfaces (i.e., models) one would like to classify. The models typically used are the six fundamental surfaces of differential geometry, i.e., plane, peak, pit, roof, valley, saddle. The 2nd phase is normally an iterative procedure aimed at refining the first segmentation. Although good results are typically obtained, they present convergency problems, so that computational complexity cannot be estimated a priori. This approach is the most common found in literature [20]-[25], and mostly used to 3-D acoustic image processing.

Hybrid methods combine the search for edge points with the determination of actual regions to get a true segmentation of an image [24,26,27].

Region-based methods are more accurate, but present higher complexity with respect to the other approaches. Among the region-based algorithms, surface fitting approaches have shown better performances, but, to have a large representational accuracy, a high degree of the polynomial should be used, at the expense of the computational cost.

Past work in acoustical imaging was concerned mainly with restoration and segmentation of acoustic data, typically acquired by sidescan sonar. Speckle noise reduction from sidescan sonar images was addressed by using Simulated Annealing and Iterative Conditional Modes [28]. Image restoration is performed assuming that speckle noise can be modelled by Rayleigh or Gaussian probability density function. The Graduated Non-Convexity [29] minimization method has been used for the reconstruction of sea bed bathymetric data acquired by a multibeam sonar [30]. Specifically, evaluation of the error of the reconstruction process is performed using a weak membrane or a thin-plate model [31] together with a line process, included to remove the smoothness constraint at surface boundaries. MRF segmentation of images acquired by a multibeam echosounder is also addressed by Dugelay *et al.* [32]. By arranging the acoustical map on the basis of the acquisition parameters, they devise a classical MRF process modelling the prior energy as weighting coefficients depending on its geometrical location and the observation energy term as a χ^2 distribution. The segmentation of simulated high frequency 3-D sonar images was presented by Subramaniam and Bahl [33]. This work is essentially a surface fitting procedure [25], used to recover 3-D surface structures from sparse data. Mignotte *et al.* [34] address the segmentation of sonar images by using an unsupervised hierarchical MRF algorithm. Sonar images of the sea bed are segmented in 2 classes, shadow and seabed reverberation, useful for detection and classification of entities lying on the seabed. Multiresolution MRF models are devised at each scale: the model is hierarchical as local neighbours are devised at each resolution level and between levels. Gaussian and Rayleigh laws are assumed for the two classes and used in the model parameters' estimation phase performed during the segmentation process without using prior knowledge. Calder *et al.* [35] present another MRF-based technique for the segmentation of sidescan sonar images. The approach here is to segment an image on the basis of its textural content. The adopted MRF models are the auto-binomial model and the multinomial model [36], also including a method for the estimation of the model parameters. Carmichael *et al.* [37] perform segmentation and classification of sidescan sonar images by using fractal measures of roughness of sediments. A multiresolution directional operator is proposed which is capable of discriminating between several types of seafloor sediments having different acoustic scattering. Locally estimated fractal measures provide a set of discriminant features that can be easily separated by typical classification algorithms. Work of Linnet *et al.* [38] presents another texture-based approach to segment sidescan images by using fractal and spatial-point approaches. The former method utilizes a maximum likelihood classificator trained with the fractal dimension of characteristic sample textures, which is directly linked to local grey-level distribution. The latter method analyses the spatial distributions of the grey levels to characterize each image pixel, still utilizing the same training samples' set. Since objects are regarded as a perturbation over the background texture, they are detected just as small unclassified image areas sunk in a larger classified zone, practically, building an associated probability map indicating the reliability of each pixel to belong to a certain texture.

Zerr and Stage present a method to estimate the 3-D structure of an object from several sonar images [39]. The method assumes that a sector-scan sonar acquires a set of images around the object of interest. Then, classification of the images in 3 types of regions (echo, shadow, background) is performed by using a couple of thresholds fixed on the basis of the sonar performances known a-priori. Therefore, shadow information in an image is used to compute the cross section of the object from that direction, and then the whole 3-D structure is obtained by volumetric reconstruction exploiting occluding contours, by combining all the estimated cross sections. The reflectivity map is estimated by computerized tomography of the echo information by using the 2D Radon transform. Unfortunately, this approach assumes that the sonar images were taken at a constant distance from the object and a constant height from the seabed, so that it can be affected by positional uncertainties typical of underwater scenarios. Works of Auran *et al.* [40] addressed the reconstruction problem from sonar data. These works utilize the occupancy grid approach [41] to support several types of AUV tasks. More specifically, sector scan data are processed to build a dynamic 3-D occupancy map where some useful information is stored by using an adequate data structure representation. This map is actually a volumetric representation in polar coordinates where present cells denote the presence of an echo at that location and a set of associated information useful for subsequent processing. After the map formation, clusters are detected by the connected component analysis in order to identify the main objects present in the scene, and useful features (i.e., moments, radial and angular sizes, bounding box, area, volume, etc.) are extracted from the clusters. The subsequent steps aim at modelling and visualizing the detected objects by a surface fitting procedure and, if different views are available, a complete reconstruction can be performed after individual surface estimation by merging surface patches.

From the above summary, one can notice that several works address the problem of segmentation or reconstruction of acoustic images using statistical or geometrical approaches, but only a few of them exploit different information sources directly provided by the sensor.

The problem of reconstruction of acoustic data extracted by an acoustical multibeam system was already addressed by the authors in [42]-[45]. In [42], a first formulation of the energy function was devised, simply assuming reliable those 3-D measures associated with high echo returns (high confidence) and discarding the low confidence range points. In [43], this mechanism is evolved in the definition of a complex functional which takes into account the physical significance of the coupling term between 3-D and confidence images. Another paper [45] presents a generalization of the latter work where theoretical justifications in probabilistic terms are provided for the functional form proposed in [43] as well as for the other novel energy formulations.

A novel approach has been recently investigated in the field of acoustic image reconstruction. For example, in survey missions, autonomous or remotely operated vehicles (AUVs, ROVs) are typically equipped by acoustic devices sensing the environment. During vehicle navigation acoustic 3-D images are acquired, but, due to the limited field of view and the implicit inaccuracy of its positioning, it is difficult to obtain a complete and accurate investigation of the surrounding environment. For these reasons, in [46] 3-D acoustic image mosaicing has been addressed. Image mosaicing is a recent technique used to align optical (2-D) images in order to reconstruct a panoramic view of the environment. In this context, the term has been used to name a 3-D reconstruction of the environment from multiple 3-D acoustic images.

This work should be considered in the research line devoted to the registration of image pairs or the integration of a set of range images. Typically, the term *registration* is used for the geometric alignment of a couple or more of 3-D data point sets, while the term *fusion* is utilized when one would like to get a single surface representation from registered 3-D data sets. Most of the work present in the literature assume range images from a laser range finder looking at a single, even complex, object.

Among these ones, works related to registration, the Iterative Closest Point (ICP) procedure [47,48] and its variants [49] are seminal papers worth to be mentioned. They are all based on the original iterative algorithm based on the search of pairs of nearest points in the two sets, and estimating the rigid transformation which align them. Then, the rigid transformation is applied to the points of one set, and the procedure is iterated until convergence is reached. Variants include the use of closest points in the direction of the local surface normal [48], and the use of a robust statistics technique [49] to limit the maximum distance between points. Originally, these works assume that one point set is a subset of the other and in not too distant positions. In this manner, several approaches have been investigated, and extended to register multiple 3-D images [50]-[55]. The underwater environment implies dealing with uncertain low resolution data, in which problems of filtering, segmentation and reconstruction should be all considered in order to get a reliable reconstruction of the scene. The work in [46] is the only one dealing with acoustic data and aims at reconstructing a 3-D environment from a sequence of clutter, noisy, and low resolution data, in order to produce a 3-D panoramic mosaic of the scene.

7. SEGMENTATION AND RECONSTRUCTION OF UNDERWATER TUBULAR STRUCTURES

In this section, we address the use of the acoustic image processing to design a complete system for segmentation, reconstruction, and final augmented rendering, of underwater tubular structures.

The acoustic camera utilized [56] is formed by a two-dimensional array of transducers sensitive to signals backscattered from the scene previously insonified by a high-frequency acoustic pulse. The whole set of raw signals is then processed to estimate signals coming from fixed steering directions (beamsignals) while attenuating those coming from other directions. Assuming that beamsignals represent the responses of a scene from a 2D set of (steering) directions, a 3-D point set can be extracted detecting the time instant t^* at which the maximum peak occurs in each beamsignal. Besides, the intensity of the maximum peak can be used to generate another image, registered with the former, representing the reliability of the associated 3-D measures. In other words, the higher the intensity, the safer the 3-D measure associated. Images are formed by 64×64 3-D points ordered according to an angular relation, as adjacent points correspond to adjacent beam signals. Their coordinates are expressed in a 3-D reference frame attached to the sensor.

7.1 3-D data processing

A preliminary low level processing phase is performed on the raw data obtained by the acoustic camera, both to clean the images from noise and to extract the 3-D edges of the objects observed.

Noise Filtering

Since acoustic raw images are typically quite noisy, due to both the environment conditions and the physical limitations of the camera, it is mandatory that a pre-processing phase be applied to reduce spurious information from such images. The used acoustic camera directly performs a preliminary low level stage: it provides together with the 3-D measures, the related intensity and other useful information associated to each point as well, like a connection matrix and normals. More specifically, data are arranged in a 64×64 array $\mathbf{C}_{ij}$ of 3-D coordinates, estimated from the time instants at which the main peak of the beamsignals are detected. An intensity value is also associated to each 3-D point, representing the reliability of the peak of the detected responses. Consequently, adjacent points in the image correspond to neighboring beamsignals, and connected components are formed by neighbouring points whose Euclidean distance is below a certain threshold T_r, set on the basis of sensor resolution and a-priori knowledge of the scene.

Using this procedure, we define two points as connected if it is possible to find a chain of neighbouring points connecting them. In such a way, it is possible to subdivide the image in a certain number of connected components, while discarding those components formed by a small number of points, likely not representing interesting physical objects.

This technique is usually called "size" filtering in the two-dimensional image processing literature. Finally, "reliable" connected components are formed by the points whose associated intensity is above a certain threshold, still depending on the camera properties.

Smoothing by line fitting

Smoothing of range data can typically be performed by locally fitting a parametric surface to range data. However, many methods based on such technique produce inaccurate results when surface or derivative discontinuities are present, and even on smooth surfaces, whenever the image contains scattered impulse values, called outliers. Thus, robust methods are needed, that are powerful enough to handle data coming from discontinuous (piecewise-smooth) surfaces and affected by different kinds of noise. When two dimensional windows are used, points within a window close to surface discontinuities come from at least two different populations: "drawing the line" between the two populations is in general a difficult problem. When more than two populations are considered (e.g., close to object corners) or non-straight edges cross the window, the problem is even harder. In order to overcome this problem, following [46,57], we use one-dimensional (linear) windows swept along several directions on the image plane, and integrate the results obtained through this first directional processing step for obtaining a solution to the two-dimensional problem. This algorithm solves the surface fitting problem on range data in a fast, highly-parallel, efficient, yet simple and robust way. Although it is suited for the special case of piecewise-linear surfaces, the method can be generalized to polynomial surfaces of higher degree.

The method consists of two distinct and independent steps. First, an isotropic set of directions is taken, and slices are extracted from the original image, in such a way that every pixel belongs to exactly one slice per direction. Each slice is then viewed as the discrete, noisy version of a piecewise-smooth function of one variable. A one-dimensional fitting algorithm is applied to each slice; the one-dimensional processing for every direction gives an estimate for the z value at each point. In the second step, all estimates obtained through the one-dimensional algorithm are considered. Different estimates of the position of one pixel are averaged in order to obtain a final estimate.

As for the first step, the method locally solves the problem by allowing the neighbourhood of each pixel to "float around", looking for a homogeneous set of data, i.e. a set which does not contain any discontinuity. The method is based on the assumption that at least one partial neighbourhood per pixel always exists whose data are homogeneous. For every point P, the algorithm fits one line per window on all windows containing P, basing on Haralick's facet model. A goodness-of-fit measure is computed for each window, then the "best" result is chosen to give estimates for the value of the underlying function at point P. We have experimented with various algorithms and goodness-of-fit measures. Under mild assumptions on the nature of the noise, a least-square algorithm and a χ^2 measure can be successfully applied, provided that data are pre-processed. This solution achieves a good compromise between accuracy and computational complexity.

Experiments and results

Several experiments have been performed to test pre-processing phase of the acoustic images, and we obtained good results both in the cleaning of the images and the general understandability of them.

In Fig. 4, a typical raw image acquired by the acoustic camera and related the filtered data are shown. Such images have been obtained using $T_r = 20$ cm and the threshold on the intensity equal to 5000 (set on the basis of the histogram distribution). We can see that the procedure is quite good and the cleaned images have been used in all the subsequent step of the recognition module.

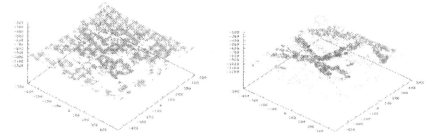

Figure 4. Raw 3-D data (left) and filtered 3-D data (right).

7.2 Image segmentation and reconstruction

After the pre-processing of the acoustic images, we obtain an ensemble of 3-D points representing, with a high confidence, physical points derived from objects in the scene. The main problem is now to segment such ensemble in well distinguished regions corresponding to different entities present in the observed scene. Since the underwater scene is composed by a structure of tubular elements, we consider mainly two kind of objects, i.e., pipe-like and not pipe-like ones.

If we are interested in the construction of an augmented representation of the underwater structure the segmentation phase to identify the pipes may suffice. Conversely, if we would like to build a virtual representation, a recognition of the joints formed by pipes intersection is necessary in order to identify them by applying a matching phase using a model reference structure. Moreover, such a matching phase is necessary for the pose and position determination of the ROV from the acoustic data interpretation.

Segmentation

Three possible techniques have been investigated to segment the acoustic images. The first one is based on a region growing algorithm guided by the fitting of the 3-D points with a set of quadrics. The second one is a morphological approach useful in finding out rectangular shapes in the range images. The third one is based on the skeleton extraction from the distribution of points and a subsequent analysis of its branches. We briefly describe these methods with emphasis on the skeleton method that is the one used for the final tests.

Geometrical method
The geometrical method, that has been extensively discussed in [58], have been tested both on real and synthetic acoustic images. It consists in the following steps:
- Points Labelling (fitting with expansion)
  1. A point with its 25 neighbours points is randomly selected to form the set R_i
  2. Points of R_i are fitted with a quadric, if fit error is too large choose another point (go to 1)
  3. Add to R_i all of the points (expansions) with error less than a threshold and go to 2; else, if there are no points to be added, start with a new region (i=i+1 and go to 1)
  4. Repeat until all points are labelled
- Refinement
  1. Points not labelled are assigned to the closest region
  2. Elimination (fusion) of smaller regions
  3. Fit repeated in all regions
- Merging of regions
  1. All pairs of regions are considered
  2. Surface of fit of the union region is computed
  3. We merge regions comparing the error of such a fit with the mean error of the fit of the separate regions; if it is less the two regions are merged.
- Labelling of isolated points
  1. Detecting isolated points using a compactness measure

2. Reassignment of these points to the nearest region that contains them.

This method is very reliable for synthetic images and, in general, for clean images where the curvature of the objects (pipes) is evident; unfortunately this is not the case for real images captured with the acoustic camera. The geometrical algorithm, in these cases, tends to under-segment the image and to fit pipes with planes. Nevertheless it remains a valid method for sharp images and an interesting one to be studied for further development.

In Fig. 5, two examples of 3-D data as obtained from the filtering stage, and segmented data are displayed. One can notice a certain degree of error in the resulting segmentation.

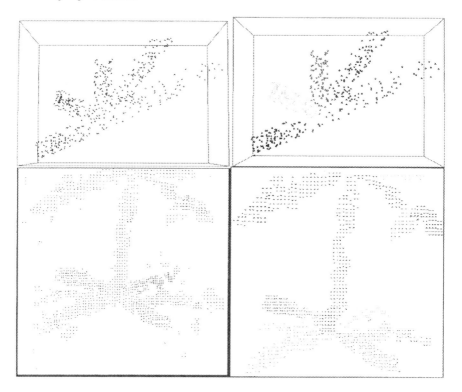

Figure 5. Two examples of filtered 3-D data (left column), and segmentation results using the geometric method (right column).

Morphological method

The second method starts from the idea that the projection of pipes on a plane gives mostly rectangular shapes. The search of pipe-like structures in the observed scene is then reduced to the search for rectangular shapes in the angular range images. The decomposition of an image in elementary shapes is a well known problem of the mathematical morphology [59]. Using standard morphological techniques we have studied the possibility to decompose the range images in elemental rectangular shapes and the subsequent back projection of these shapes in the 3-D space. Although the results are very good in some cases, this method is not robust to noise and tends to over-segment the real images; furthermore, morphological operators, in general, tends to be very CPU intensive procedures and so are not very suitable for a real time application.

In Fig. 6, a couple of examples of 3-D data as obtained from the filtering stage, and segmented data are displayed. A certain degree of over-segmentation is evident, resulting in decomposing single pipes in more regions.

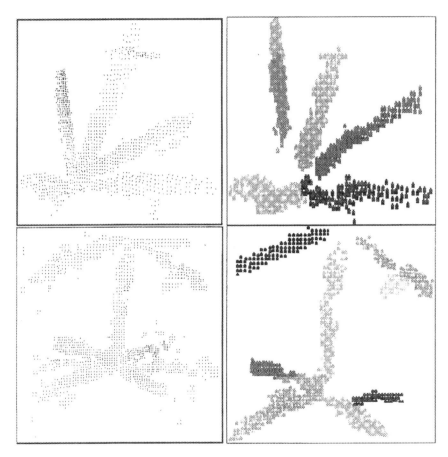

Figure 6. Two examples of filtered 3-D data (left column), and segmentation results using the morphological method (right column).

Skeleton-based method

The third algorithm is very simple and customized to deal with the extraction of tubular structures, even if it is not prevented its application to different kind of objects. A 3-D skeleton is first extracted and then used to subdivide the image in different convex components. Since this method is quite robust to noise and gives a good segmentation with respect to the other two methods, although it is possible to improve it in many ways, it has been chosen as the one utilized for the final tests and experiments [60].

To extract the skeleton, we apply to the image the following procedure: for every point A in the image we consider all the points that are in a sphere of radius R centred on A. We then shift A from its actual position to the centre of mass of such distribution of points. We apply this procedure in a parallel way on all the points of the image.

In short, the skeleton extraction is performed along the following lines. Let us define the 3-D image as I, i.e., the ensemble of points in $\mathbf{R}^3$.

$$I = \{\mathbf{x}_i\} = \{x_i, y_i, z_i\}, \quad i = 1,\ldots,N \tag{16}$$

We also define for each point i and each R in $\mathbf{R}^+$ (ray of a sphere about the point i) the subset O_i^R of I and the 3-D point $\mathbf{b}_i^R$ defined in the following way:

$$O_i^R \equiv \{\mathbf{x}_j \in I : |\mathbf{x}_i - \mathbf{x}_j| < R\} \tag{17}$$

$$\mathbf{b}_i^R \equiv \frac{\sum_{\mathbf{x}_j \in O_i^R} \mathbf{x}_j}{\dim\{O_i^R\}} \tag{18}$$

Let us define the *interior* λ_i^R for the point i in the following way:

$$\lambda_i^R \equiv \frac{R - |\mathbf{x}_j - \mathbf{b}_i^R|}{R} \tag{19}$$

One can easily verify that:

$$0 \leq \lambda_i^R \leq 1 \quad \forall i \tag{20}$$

The interior is a measure of how much a point is "inside" the object specified by I. If $\lambda_i^R \approx 1$, the point is inside a homogeneous zone of radius at least equal to R; on the contrary, for $\lambda_i^R \approx 0$, the point is very near to a border of the three-dimensional distribution of points. Defining the following image transformation:

$$I \rightarrow I^R = \{\mathbf{b}_i^R\} \tag{21}$$

and indicating with the symbol $I^{R,n}$ the iterative application of it for n times, our skeleton extraction is simply the construction of the image $I^{R,n}$ for a suitable choice of R and n. The overall effect of this transformation is to shift points on the border, i.e., points with a low λ^R, toward the centre while leaving points well inside an object, i.e., points with a almost unitary value of λ^R, unaltered. The iterative application of such a procedure tends to shift all the points of the distribution towards the skeleton. Such an algorithm may be seen from a physical point of view like a short range interaction between physical points.

The choice of the free parameter R is very important. If it is too small, with respect to the average mutual distances of the points, the distribution would likely collapse in a certain number of disconnected punctual regions, preventing the extraction of the skeleton. If it is too large the value of λ may be small also for points well inside the object and again the skeleton will not be properly extracted. The parameter R plays a similar role played by the dimension of the structuring element in the classical mathematical morphology. It is straightforward to verify that the skeleton so extracted is invariant for three dimensional rotation (condition that is never exactly satisfied for skeletons of a 2D bitmap image. Preliminary experimental results show also that the homotopy group is preserved by the algorithm. Using a priori knowledge of the structure represented by the 3-D image and of the sensor resolution, it is possible to estimate proper values for R and n aimed at improving the extraction of a good skeleton. In the results, R = 70 and n = 2 are used.

A possible improvement for this algorithm is to find a procedure that is automatically able to set the free parameter R depending on 3-D distribution characteristics (like points density, moments, etc.). Actually, such selection should be local, so that the skeleton will be extracted in different ways in different image regions, so that the method could be applied to any kind of data without prior knowledge of the scene investigated. Moreover, we are also planning the implementation of an estimator to evaluate the "quality" of the skeleton obtained, in order to have an automatic criterion to stop to the algorithm. Several ideas are currently investigated to solve these two problems. We suppose now that every branch of the skeleton corresponds to a component object in the scene. Therefore, the segmentation is obtained by classifying every point of the skeleton in either *branch point* or *joint point*. Such a classification depends on the neighbourhood around the point: if all the neighbours (e.g., the ones contained in a small sphere) belong to a quite straight segment, it is classified as a branch point, otherwise it is a joint point. If we now consider only the branch points, we obtain well separated connected components of the skeleton; we can segment them with a percolation technique and associate to each of them an object of the scene. In other words we can search for the connected components of the skeleton minus its joints in the same way we obtained the connected components for the 3-D distribution of points in the pre-processing phase; these regions forms the branches of the skeleton and are the candidates to be classified as pipes.

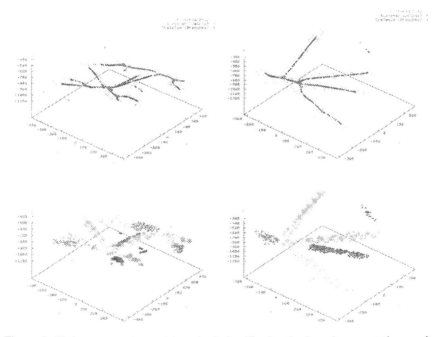

Figure 7. Skeleton extraction and branches' classification (top), and segmentation results using the skeleton-based method (bottom).

In Fig. 7, the results of the skeleton extraction algorithm are presented related to two 3-D images of a tubular structure (top), and the resulting segmented images are shown in the bottom part of the same figure. This method, although not perfect, appears more accurate and robust with respect to the above-mentioned approaches.

Classification

The segmentation procedure does not yet classify the regions in pipe-like and not-pipe-like elements. To perform this last step, we need to analyse the 3-D geometric shape of these regions. To obtain information on the shape of a 3-D distribution of points, it is possible to use a particular tensor associated to such distribution, the Inertial Tensor; such technique is deeply related, as we will show, to the so called Principal Component Analysis.

If ρ is a continuous distribution in R^3, we define the Inertial Tensor I of ρ as a 3x3 matrix given by the following formulas:

$$IT_{xx} = \int \rho(x,y,z)(y^2 + z^2)dV$$
$$IT_{yy} = \int \rho(x,y,z)(x^2 + z^2)dV$$
$$IT_{zz} = \int \rho(x,y,z)(x^2 + y^2)dV \qquad (22)$$
$$IT_{xy} = IT_{yx} = -\int \rho(x,y,z)\,xy\,dV$$
$$IT_{yz} = IT_{zy} = -\int \rho(x,y,z)\,yz\,dV$$
$$IT_{xz} = IT_{zx} = -\int \rho(x,y,z)\,xz\,dV$$

where the coordinates are set so that the centre of mass of the distribution is in the origin. The eigenvalues of this tensor corresponds to the rotation inertia of the distribution around the principal axis defined by the directions of the corresponding eigenvectors of I.

For a discrete distribution we indicate with **o** the centre of mass of the distribution:

$$\mathbf{o} = \frac{\sum_i \mathbf{b}_i}{N} \quad i \in C_k \qquad (23)$$

where $\mathbf{b}_i$ are the position vectors of the distribution (the apex R is omitted for simplicity), and C_k is the k-th connected component. For the inertial tensor we have:

$$IT = \sum_i (\mathbf{b}_i - \mathbf{o}) \Theta (\mathbf{b}_i - \mathbf{o}) \qquad (24)$$

where we use the symbol Θ for the following operator:

$$\mathbf{a}\Theta\mathbf{b} = \begin{bmatrix} (a_z b_z + a_y b_y) & -a_x b_y & -a_x b_z \\ -a_y b_x & (a_z b_z + a_x b_x) & -a_y b_z \\ -a_z b_x & -a_z b_y & (a_y b_y + a_x b_x) \end{bmatrix} \qquad (25)$$

In the case of a symmetric distribution, the eigenvectors of such a matrix, being the principal inertial axes, are symmetry axes too. For example, for a pipe-like distribution, one of the eigenvector, the one corresponding to the lowest eigenvalue, lies along the pipe axis. In this way, we can use eigenvalues and eigenvectors of I to extract useful information on the shape of the discrete distribution.

The Principal Component Matrix [60] is essentially defined as:
$$PCM = \sum_i (b_i - o) \otimes (b_i - o) \qquad (26)$$
where the symbol $\otimes$ stands for the following operation:
$$a \otimes b = \begin{bmatrix} a_x b_x & a_x b_y & a_x b_z \\ a_y b_x & a_y b_y & a_y b_z \\ a_z b_x & a_z b_y & a_z b_z \end{bmatrix} \qquad (27)$$
One can see that from the preceding definitions we have:
$$PCM + IT = \left[\sum_i (b_i - o) \cdot (b_i - o) \right] \cdot \mathbf{1} = \alpha \cdot \mathbf{1} \qquad (28)$$
where $\mathbf{1}$ is the identity matrix. From this relation it follows that every eigenvector v of PCM with eigenvalue λ is also an eigenvector of I with eigenvalue $(\lambda-\alpha)$. In fact:
$$PCM \cdot v = \lambda \cdot v. \qquad (29)$$
Hence:
$$(PCM + IT) \cdot v = \lambda \cdot v + IT \cdot v = \alpha \cdot v \qquad (30)$$
from which:
$$IT \cdot v = (\alpha - \lambda) \cdot v. \qquad (31)$$

Consequently, it is clear that one can use both methods to obtain the same results. We have chosen to use the Inertial Tensor: we calculate I for every segmented region in the scene and obtain three eigenvalues. If one is smaller with respect to the others, and these are of the same magnitude, we classify the region as a tubular one. Obviously, we have also to introduce a threshold: if it is too small, it is probable to classify as a pipe something that is only elongated, whereas, if it is too high, it is probable to lose some pipes from the scene.

$$\begin{aligned} & e_j \quad j = 1,...,3 \\ & e_m \ll e_k \quad m \neq k \\ & e_n \cong e_h \quad n, h \neq m \end{aligned} \qquad (32)$$

From the value of the minimum eigenvalue it is possible to extract also the order of magnitude of the radius of the tubular region. In fact, in the case of a complete cylindrical distribution the following relation holds:
$$e_m = 0.5 \cdot M \cdot r^2 \qquad (33)$$
where e_m is the minimum eigenvalue, M is the total number of points in the distribution and r is the radius. As in the range images of underwater pipes, data are not distributed on the surface of a cylinder, but only on a little portion of it (the portion "visible" from the camera), this relation is only approximately true, but it is sufficient to give the right order of magnitude for the radius, as we will see in the following. For a more precise extrapolation of this parameter, it is necessary a more sophisticated algorithm that is based on a robust fitting procedure. Another problem that is actually addressed in our research work is a fast and reliable refinement of the segmentation. Because of noise and the particular skeleton reconstruction procedure, it is quite probable to have an oversegmentation of the image, i.e. a single object may be segmented in more than one region. This problem is partially solved considering every pair of regions and calculating the inertia tensor of their union: if it is "more" pipe-shaped (by using the previous criterion), the two regions are merged.

We have applied our classification algorithm to the segmented images previously obtained and we have extracted the pipes as shown in Fig. 8. From these examples, it is clear the necessity, at least for Joint 1 (Fig. 7, bottom-left), of the merging phase sketched above; in fact some of the pipes in this scene have not been recognized because they were oversegmented in the classification step. This problem will be solved after the merging phase will be completed. It should be noted, however, that the Joint 1 is quite an extreme case that we present here only for completeness; in most of the underwater images we have analysed the results are good and similar to the ones obtained for Joint 2 (Fig. 7, bottom-right).

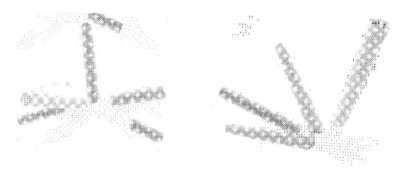

Figure 8. Augmented reality examples.

It is possible now to proceed in two different, but not excluding, ways: an augmented reality or a complete virtual reality representation [61]. The first one is simply obtained by projecting the virtual pipes, corresponding to the 3-D segmented regions, on the image plane. The resulting image is more easily readable by a human operator, and hence, useful for navigation tasks or simply to realize what the camera is looking at. This procedure does not require any prior knowledge of the structure investigated (apart the generic knowledge on the objects' shape). As an example, we have rendered a scene with the data (raw and filtered) of the underwater acoustic images together with the pipes extracted with our algorithm (see Fig. 8); it is interesting to note that the radius of the pipes are quite good although obtained with the approximated formula.

The complete virtual environment requires a preliminary recognition phase. The idea is to build, from a database, a virtual scene of the structure investigated (e.g., an off-shore platform) and to use the segmented images to recognize the position and the pose of the acoustic camera in the scene by a suitable matching with the database information. Therefore, a human operator could not observe the real image, but the virtual one, possibly navigating in the virtual scenario reconstructed from real data.

8. CONCLUSIONS

We have presented a general overview of the approaches and methods aimed at the generation and processing of acoustic images. In particular, in the first part, methods for the generation of acoustic images are presented, describing holographic, and lens-based methods, and, in detail, the 3-D beamforming technique. In the second part, the emphasis was on 3-D image processing techniques. After a general survey about the methods devoted to range image processing, a specific overview of the methods for acoustic image analysis and segmentation is reported, and finally, a technique for 3-D acoustic image segmentation and interpretation is presented, aimed at obtaining a synthetic representation of the observed scene, useful for augmented reality and virtual reality applications.

ACKNOWLEDGEMENTS

The authors would like to thank Dr. R.K. Hansen of Omnitech A/S (Norway) for kindly providing the images acquired by the Echoscope acoustic camera, and Dr. R. Giannitrapani, who partly contributes to this project. This work was partly supported by the European Commission under the BRITE-EURAM III project no. BE-2013 VENICE (Virtual Environment Interface by Sensory Integration for Inspection and Manipulation Control in Multifunctional Underwater Vehicles).

REFERENCES

[1] A. Macovski, "Ultrasonic Imaging Using Arrays," *IEEE Proceedings*, vol. 67, pp. 484-495, April 1979.
[2] J.L. Sutton, "Underwater Acoustic Imaging," *IEEE Proceedings*, vol. 67, pp. 554-566, April 1979.
[3] V. Murino and A. Trucco, "Three-Dimensional Image Generation and Processing in Underwater Acoustic Vision," *IEEE Proceedings*, vol. 88, pp. 1903-1946, December 2000.
[4] O. George and R. Bahl, "Simulation of Backscattering of High Frequency Sound From Complex Objects and Sand Sea-Bottom," *IEEE Jour. Ocean. Engin.*, vol. 20, pp. 119-130, April 1995.
[5] T.L. Henderson and S.G. Lacker, "Seafloor Profiling by a Wideband Sonar: Simulation, Frequency-Response, Optimization, and Results of a Brief Sea Test," *IEEE Jour. Ocean. Engin.*, vol. 14, pp. 94-107, January 1989.
[6] B.D. Steinberg and H.M. Subbaram, *Microwave Imaging Technique*, J. Wiley & Sons, New York, 1991.
[7] J.W. Goodman, *Introduction to Fourier Optics*, McGraw-Hill, New York, 1968.
[8] L.J. Ziomek, "Three Necessary Conditions for the Validity of the Fresnel Phase Approximation for the Near-Field Beam Pattern of an Aperture," *IEEE Journal of Oceanic Engineering*, vol. 18, pp. 73-75, January 1993.
[9] R.O. Nielsen, *Sonar Signal Processing*, Artech House, Boston, 1991.
[10] J.S. Jaffe, P.M. Cassereau, and D.J. Glassbrenner, "Code Design and Performance Characterization for Code Multiplexed Imaging," *IEEE Trans. Acoustics, Speech, Sig. Proc.*, vol. 38, pp. 1321-1329, August 1990.
[11] J.C. Bu, C.J.M. van Ruiten, and L.F. van der Wal, "Underwater Acoustical Imaging Algorithms," *Proc. Europ. Conf. on Underwater Acoustics*, Luxembourg, pp. 717-720, September 1992.
[12] R.K. Hansen and P.A. Andersen, "3D Acoustic Camera for Underwater Imaging," in *Acoustical Imaging*, Yu Wei and Benli Gu, Eds., vol. 20, pp. 723-727, New York, 1993, Plenum Press.
[13] R.K. Hansen and P.A. Andersen, "A 3D Underwater Acoustic Camera Properties and Applications," in *Acoustical Imaging*, P. Tortoli and L. Masotti, Eds., vol. 22, 22nd Int.

Symposium, pp. 607-611, New York, 1996, Plenum Press.
[14] R.K. Hansen and P.A. Andersen, "The Application of Real Time 3D Acoustical Imaging," *IEEE/OES Int. Conf. Oceans '98*, Nice, France, pp. 738-741, September 1998.
[15] R. Hoffman and A.K. Jain, "Segmentation and Classification of Range Images," *IEEE Trans. on Pattern Analysis and Machine Intelligence*, vol. 9, n. 5, pp. 608-619, September 1987.
[16] Y. Shirai, "*Three-Dimensional Computer Vision*," Springer-Verlag, Germany, 1987.
[17] M.A. Wami and B.G. Batchelor, "Edge-Region-Based segmentation of Range Images," *IEEE Trans. on Pattern Analysis and Machine Intelligence*, vol. 16, n. 3, pp. 314-319, 1994.
[18] O.R.P. Bellon and C.L. Tozzi, "An Integrated Approach for Segmentation and Representation of Range Image," *ICAP'97*, pp. 279-286, 1997.D. Zhao, X. Zhang, "Range-Data-Based Object Surface Segmentation via Edges and Critical Points," *IEEE Trans. on Image Processing*, vol. 6, n. 6, pp. 826-820, June 1997.R.W. Taylor, M. Savini, and A.P. Reeves, "Fast Segmentation of Range Image into Planar Region," *Computer vision, graphics, and image processing*, vol. 45 pp. 42-60, 1989.
[21] P.J. Besl and R.C. Jain "Invariant Surface Characteristic for 3D Object Recognition in Range Images," *Computer vision, graphics, and image processing*, vol. 33 pp. 33-80, 1986.
[22] S.Z. Li, "Toward 3D Vision Range Images: An Optimization Framework and Parallel Networks," *CVGIP: Image Understanding*, vol. 55, n. 3, pp. 231-260, 1992.
[23] E. Trucco and R.B. Fisher "Experiments in Curvature - Based Segmentation of Range Data," *IEEE Trans. on Pattern Analysis and Machine Intelligence*, vol. 17, n. 2, pp. 177-182, 1995.
[24] N. Yokoya and M.D. Levine "Range Image Segmentation of Differential Geometry: A Hybrid Approach," *IEEE Trans. on Pattern Analysis and Machine Intelligence*, vol. 11, n. 6, pp. 643-649, June 1989.
[25] P.J. Besl and R.C. Jain, "Segmentation Through Variable-Order Surface Fitting," *IEEE Trans. on Pattern Analysis and Machine Intelligence*, vol. 10, n. 2, pp. 167-192, March 1988.
[26] S.M. Bhandarkar and A. Siebert, "Integrating Edge and Surface Information for Range Image Segmentation," *Pattern Recognition*, vol. 25, n. 9, pp. 947-962, 1992.
[27] D.P. Mital, E.K. Teoh, and A.W.T. Lim, "A Hybrid Method towards the Segmentation of Range Images for 3-D Object Recognition," *Int. Journal on Pattern Recognition and Artificial Intelligence*, vol. 8, pp. 869-995, 1994.
[28] R.S. Beattie and S.C. Elder, "Side Scan Sonar Image Restoration using Simulated Annealing and Iterative Conditional Modes," *Proc. Int. Conf. on Sonar Signal Processing*, Loughborough (UK), pp. 161-167, December 1995.
[29] A. Blake and A. Zisserman, *Visual Reconstruction*, MIT Press, Cambridge, 1987.
[30] S. Show and J. Arnold, "Automated Error Detection in Multibeam Bathymetry Data," *Proc. IEEE OCEANS '93*, Victoria (Canada), vol. II, pp. 89-94, 1993.
[31] R. Szelinsky, *Bayesian Modeling of Uncertainty in Low-Level Vision*, Kluver Academic Publ., Boston, 1989.
[32] S. Dugelay, J.M. Augustin, and C. Graffigne, "Segmentation of Multibeam Acoustic Imagery in the Exploration of the Deep Sea Bottom," *13th Int. Conf. on Pattern Recognition*, Vienna, pp. 437-445, August 1996.
[33] L.V. Subramaniam and R. Bahl, "Segmentation and Surface Fitting of Sonar Images for 3D Visualization," *Proc. 8th Int. Symp. on Unmanned Untethered Submersible Technology*, Durham (NH, USA), pp. 290-298, September 1995.
[34] M. Mignotte, C. Collet, P. Perez, and P. Bouthemy, "Unsupervised segmentation applied on sonar images," in *Energy Minimization Methods in Computer Vision and Pattern Recognition, Lecture Notes in Computer Science 1223*, pp. 491-506, 1997.
[35] B.R. Calder, L.M. Linnett, and S.J. Clarke, "Spatial Interaction Models for Sonar Image

Data," *Proc. Int. Conf. on Sonar Signal Processing*, Loughborough (UK), pp. 81-90, December 1995.
[36] J. Besag, "On the Statistical Analysis of Dirty Pictures," *Journal of the Royal Statistical Society*, vol. B-48, pp. 259-302, 1986.
[37] D.R. Carmichael, L.M. Linnett, and S.J. Clarke, "A Multiresolution Directional Operator for Sidescan Sonar Image Analysis," *Proc. Int. Conf. on Sonar Signal Processing*, Loughborough (UK), pp. 49-57, December 1995.
[38] L.M. Linnett, S.J. Clarke, and D.R. Carmichael, "The Analysis of Sidescan Sonar Images for Seabed Types and Objects," *Proc. 2nd Conf. on Underwater Acoustics*, pp. 733-738, Copenhagen, July 1994.
[39] B. Zerr and B. Stage, "Three-dimensional reconstruction of underwater objects from a sequence of sonar images," *IEEE Int. Conf. on Image Processing*, vol. 3, pp. 927-930, September 1996.
[40] P.G. Auran and K.E. Malvig, "Realtime extraction of connected components in 3d sonar range images," *IEEE Int. Conf. on Computer Vision and Pattern Recognition*, pp. 580-585, San Francisco, CA, USA, June 1996.
[41] A. Elfes, "Sonar-Based Real-World Mapping and Navigation," *IEEE Journal on Robotics and Automation*, vol. 3, n. 3, pp. 249-265, 1987.
[42] V. Murino, "Acoustic Image Reconstruction by Markov Random Fields," *Electronics Letters*, vol. 32, n. 7, pp. 697-698, 1996.
[43] V. Murino, A. Trucco, and C.S. Regazzoni, "A Probabilistic Approach to the Coupled Reconstruction and Restoration of Underwater Acoustic Images," *IEEE Trans. on Pattern Analysis and Machine Intelligence*, vol. 20, n. 1, pp. 9-22, 1998.
[44] V. Murino and A. Trucco, "Markov-based Methodology for the Restoration of Underwater Acoustic Images," *International Journal of Imaging Systems and Technology*, vol. 8, n. 4, pp. 386-395, 1997.
[45] V. Murino, "Reconstruction and Segmentation of Underwater Acoustic Images combining Confidence Information in MRF Models," *Pattern Recognition*, vol. 34, n. 5, pp. 981-997, May 2001.
[46] V. Murino, A. Fusiello, N. Iuretigh, and E. Puppo, "3D Mosaicing for Environment Reconstruction," *15th Int. Conf. on Pattern Recognition*, Barcelona, Spain, pp. 362-366, September 2000.
[47] P. Besl and N. McKay, "A method for registration of 3-D shapes," *IEEE Trans. on Pattern Analysis and Machine Intelligence*, vol. 14, n. 2, pp. 239-256, February 1992.
[48] T. Chen and G. Medioni, "Object modeling by registration of multiple range images," *Image and Vision Computing*, vol. 10, n. 3, pp. 145-155, 1992.
[49] Z. Zhang, "Iterative point matching of free-form curves and surfaces," *Int. Journal of Computer Vision*, vol. 13, n. 2, pp. 119-152, 1994.
[50] G. Blais and M.D. Levine, "Registering Multiview Range Data to Create 3D Computer Objects," *IEEE Trans. on Pattern Analysis and Machine Intelligence*, vol. 17, n. 8, pp. 540-547, 1995.
[51] R. Bergevin, M. Soucy, H. Gagnon, and D. Laurendeau, "Towards a general multiview registration technique," *IEEE Trans. on Pattern Analysis and Machine Intelligence*, vol. 18, n. 5, pp. 540-547, May 1996.
[52] A.J. Stoddart, A. Hilton, "Registration of Multiple Point Sets," *Int. Conf. on Pattern Recognition*, Vienna, pp. 40-44, 1996.
[53] K. Pulli, "Multiview Registration for Large Data Sets," *Int. Conf. on 3D Digital Imaging and Modeling*, Ottawa, pp. 160-168, 1999.
[54] A.E. Johnson, M. Hebert, "Surface Registration by Matching Oriented Points," Int. *Conf. on 3D Digital Imaging and Modeling*, Ottawa, Canada, pp. 121-128, 1997.
[55] M. Soucy and D. Laurendeau, "A General Surface Approach to the Integration of a Set of Range Views," *IEEE Trans. On Pattern Analysis and Machine Intelligence*, vol. 17, n. 4, pp. 344-358, April 1995.

[56] R.K. Hansen and P.A. Andersen, "A 3D Underwater Acoustic Camera - Properties and Applications," in *Acoustical Imaging*, P. Tortoli, L. Masotti Eds., Plenum Press, pp. 607-611, New York, 1996.

[57] E. Puppo and L. Davis, "Surface Fitting of Range Images Using a Directional Approach," *Technical Report CAR-TR-504*, Center for Automation Research, Univ. of Maryland, 1990.

[58] V. Murino, A. Grion, and S. Bianchini, "A Geometric Approach to the Segmentation and Reconstruction of Acoustic Three-Dimensional Data," *OCEANS '98 MTS/IEEE*, pp. 582-586, Nice, France, September 1998.

[59] J. Serra, *Image analysis and mathematical morphology*, Academic Press, London, 1982.

[60] V. Murino and R. Giannitrapani, "Three-Dimensional Skeleton Extraction by Point Set Contraction," *IEEE Int. Conf. on Image Processing ICIP '99*, pp. 565-569, Kobe, Japan, October 1999.

[61] R. Giannitrapani, V. Murino, and A. Trucco, "Segmentation of Underwater 3D Acoustical Images for Augmented and Virtual Reality Applications," *Int. Conf. Oceans '99 MTS/IEEE*, pp. 459-465, Seattle, WA (USA), September 1999.

Index

A

Acoustic camera 260,261,264
Acoustic communication 27,227
Acoustic lens 248,250
Adaptive algorithms 1,18-21,24,27
ADPCM (Adaptive Differential Pulse Code Modulation) 134,141
Ambient noise *see noise*
Ambient pressure 129-132
Ambiguities 47,48,55,252
Ambiguity function 52,235
ASK (Amplitude Shift Keying) 104,105,142,143
Angular ambiguity 48,49
Angular resolution 252
Array processing 6-8,11,25,50,128
Assembly programming 109
AUV (Autonomous Underwater Vehicle) 2,10,128,168,259
Auto-focusing 61-66,74-76,79,84
Average Magnitude Difference Function (AMDF)136,137,139
Axial resolution 40,41,43,83
Azimuthal resolution 40,42,43,47, 70,73,83

B

Bandwidth 4,5,97,102,128,133, 134,144
Bandwidth efficiency 9
Bathymetry map 68
Beamforming 12,57,64,128,248,250-255
Beampattern 40,53,207,215,252
Beahavioural monitoring 90
Biological scatterers 205
Biomass 205,206,210
Biotelemetry 89-120
Breathing noise 128,130
Breathing rate 115,116
Bright spot 62,63,71,74
Bubble noise 128,130,136

C

Channel *see communication channel*
Channel modeling 7
Chirps 53,76
CELP (Code Excited Linear Prediction) 135,141,145
Coding 9,28,103,227,228,130
Coherentmodulation/detection1,7, 10,142
Communication channel 1,4,14,19,24, 28,101,145,232,234
Compression 3,27,130,134
Contrast function 61
Control 1,3
Correlation sonar 37,45,46

D

De-emphasis 141
Deep sound channel 6
Delay Estimation 46
Delay time difference *see Time delay difference*
Delta Modulation 134
Displaced phase centers 64
Divers 2,90,94,95,110,113,116,118, 127-150
Donald Duck effect 129
Doppler 7,15,45,49,51-53,235
DOA estimation 168,190-200
DSP 19,25,26,90,91,119,129-132, 144,147,168,176-183,239
Dynamic focusing 47,54

E

Earphones 129
Echo counting 205
Echo ensembles 205-223
Echo integration 210
Echoes 42-44,248,259
Echogram 43
Echo-sounding 205
ECG (Electro Cardio Gram) 94,95,105-107,110-117

Edge-based method 257
Electromyogram 92
Encoding *see coding*
ESPRIT 194,195
Excitation mode 136,137

F

Fish farming 93
FM 92,95,230,231
Formants 135-137
FPGA 178
Fresnel approximation 251,255
FSK 9,10,26,104,105,142-144,225, 229,244

G

Gain 139,140
Gauss-Seidel relaxation algorithm 183
Geometrical method 263

H

Heart rate 106,115-118
Helium speech 129
Holographic method 250,251,255
Hopfield network 168,171,173,180, 182,196-200
Hyperbolilc law 51

I

Image 3,11,27,57,58,81
Image classigication 268
Imaging 37-85,247-271
Image processing 257
Image reconstruction 248,260,263
Image representation 256,257
Image segmentation 248,257,258, 260,263
Input/output equation 170,171
Intelligibility 132
Interpolation 233
Intersymbol interference (ISI) 5,7,8, 11,13,20,23,28,128,143,228,229
Iterative closest point 260

L

Lambert's law 38,39
Learning *see training*
Least Mean Squares (LMS) 11,19
Least Squares (LS) 184,186,200
Linear predictive coding (LPC) 3,135-150
Lyapunov energy function 198

M

Mapping 2,37-85,259
Marine animals 92
Mean squared error (MSE) 18,23
Micro-controller 91,106-119
Microphone 129
Minimum-norm solution 265
Modem 26
Modulation/detection 103,130
Mono-static sounding 206-212,220
Monte Carlo simulation 205,209,212-215
Morphological method 264
Mosaicing 259
Multi-layer perceptron (MLP) 171
Multiple-access communications 24,25,29
Multipath 1,4-6,12,15,91,95-97,101-105,128,134,143,144,203,228,231, 234,241,244
MUSIC 194,195,221

N

Navigation...3,77,84
Networks...1,2,25,27,29
Network topology...170-174
Neural networks...118,167-203
Neural network computer (NNC)...176-179
Neuron...169,170
Noise...4,56,231,234,258,261
Noncoherent modulation/detection 1,7,8,142

O

OOK 105-107
Optical imaging 247

P

Parametric coding 137
Pattern Time Delay Shift Coding (PDS) 227-245
PCM (Pulse Code Modulation) 103,115-117,134,141
PDF method 216-218
Perceptron 171-173
Phase coherent *see coherent*
Phase Locked Loop (PLL) 15,18
Piezo-ceramic speakers 132
Pitch period 135-138
Pre-empasis 138
Processing element 176,177,184, 187,190
Press to talk 132
PPM (Pulse Position Modulation) 103,116,117,133,134,143-149
PSK (Phase Shift Keying) 10-12,16,23,26,28,104,142,144
Psychological monitoring 90
PWM (Pulse Width Modulation) 103

Q

QAM (Quadrature Amplitude Modulation) 10,12,16,28
Q factor 102,145
QPSK *see PSK*
Quantization 139-149,232

R

Radial basis function (RBF) network 168,172,184,187
Radio telemetry 93,103
Raised cosine pulse 15,22
Range 4,5
Range ambiguity 47
Range resolution 253
Ranked array method 216-218,221
Rayleigh fading 7
Rayleigh scattering 211
Recursive Least Squares (RLS) 12, 20,22,23
Recurrent neural network 173
Reduced-complexity processing 21, 27

Region-based method 257
Reflection coefficients 139,140
Resolution 61,66,79,80
Resolution cell 253-256
Respiration rate 106

S

Scan conversion 256
Shadows 42-44,73
Shadowing 210
Side scan sonar 37-45,48,54,83,258
Simulated annealing 258
Signal to noise ratio (SNR) 4,23,56, 70,100,104,133,238
Skeleton-based method 265
Sonar equations 98
Speech 2,3,27,127-150
Speech reconstruction 140
Speech synthesis 140-142
Spread spectrum 11
Surface fitting 262
Synapses 169
Synchronization 12-15,104,144,148, 231,239
Synthetic aperture sonar (SAS) 37,48-55,64,73,82
Synthetic array 53,54
Systolic array 184-190

T

Target detection 170
Target reconstruction 73
Target strength 215-223
Telemetry 3,9,90,91,96,101
Time delay difference 228,230,232-235,240
Time variation 6,12,14
Training 19,170,174-176,186
Trajectory disturbance 57-61,74
Transmission loss 4,99

U

Ultrasonic telemetry 92,93
Unvoiced *see voiced*
UUV *see AUV*

V

Vernier processing 66
Video 3
VLSI array processor 177,178,184-190
Vocal tract 136,139
Voice communications 127-150
Voiced/unvoiced sound 137,139,140
Voxel 256

W

Waveform coding 134

Printed in the United States
115602LV00003B/65/A